Advances in Intelligent Systems and Computing

Volume 995

The series "Advances in Intelligent Systems and Computing" contains publications on theory, applications, and design methods of Intelligent Systems and Intelligent Computing. Virtually all disciplines such as engineering, natural sciences, computer and information science, ICT, economics, business, e-commerce, environment, healthcare, life science are covered. The list of topics spans all the areas of modern intelligent systems and computing such as: computational intelligence, soft computing including neural networks, fuzzy systems, evolutionary computing and the fusion of these paradigms, social intelligence, ambient intelligence, computational neuroscience, artificial life, virtual worlds and society, cognitive science and systems, Perception and Vision, DNA and immune based systems, self-organizing and adaptive systems, e-Learning and teaching, human-centered and human-centric computing, recommender systems, intelligent control, robotics and mechatronics including human-machine teaming, knowledge-based paradigms, learning paradigms, machine ethics, intelligent data analysis, knowledge management, intelligent agents, intelligent decision making and support, intelligent network security, trust management, interactive entertainment, Web intelligence and multimedia.

The publications within "Advances in Intelligent Systems and Computing" are primarily proceedings of important conferences, symposia and congresses. They cover significant recent developments in the field, both of a foundational and applicable character. An important characteristic feature of the series is the short publication time and world-wide distribution. This permits a rapid and broad dissemination of research results.

** Indexing: The books of this series are submitted to ISI Proceedings, EI-Compendex, DBLP, SCOPUS, Google Scholar and Springerlink **

More information about this series at http://www.springer.com/series/11156

Rituparna Chaki · Agostino Cortesi ·
Khalid Saeed · Nabendu Chaki
Editors

Advanced Computing and Systems for Security

Volume Nine

 Springer

Editors
Rituparna Chaki
A. K. Choudhury School
of Information Technology
University of Calcutta
Kolkata, West Bengal, India

Khalid Saeed
Faculty of Computer Science
Bialystok University of Technology
Bialystok, Poland

Agostino Cortesi
Full Professor of Computer Science
DAIS—Università Ca' Foscari
Venice, Venezia, Italy

Nabendu Chaki
Department of Computer Science
and Engineering
University of Calcutta
Kolkata, West Bengal, India

ISSN 2194-5357   ISSN 2194-5365   (electronic)
Advances in Intelligent Systems and Computing
ISBN 978-981-13-8961-0   ISBN 978-981-13-8962-7   (eBook)
https://doi.org/10.1007/978-981-13-8962-7

This Springer imprint is published by the registered company Springer Nature Singapore Pte Ltd.
The registered company address is: 152 Beach Road, #21-01/04 Gateway East, Singapore 189721, Singapore

Preface

This volume contains the revised and improved version of papers presented at the 6th International Doctoral Symposium on Applied Computation and Security Systems (ACSS 2019) which took place in Kolkata, India, during 12–13 March 2019. The University of Calcutta in collaboration with Ca' Foscari University of Venice, Italy, and Bialystok University of Technology, Poland, organized the symposium. This symposium is unique in its characteristic of providing Ph.D. scholars an opportunity to share the preliminary results of their work in an international context and be actively supported towards their first publication in a scientific volume.

In our pursuit of continuous excellence, we aim to include the emergent research domains in the scope of the symposium each year. This helps ACSS to stay in tune with the evolving research trends. The sixth year of the symposium was marked with a significant improvement in overall quality of papers, besides some very interesting papers in the domain of security and software engineering. We are grateful to the Programme Committee members for sharing their expertise and taking time off from their busy schedule to complete the review of the papers with utmost sincerity. The reviewers have pointed out the improvement areas for each paper they reviewed, and we believe that these suggestions would go a long way in improving the overall quality of research among the scholars. We have invited eminent researchers from academia and industry to chair the sessions which matched their research interests. As in previous years, the session chairs for each session had a prior go-through of each paper to be presented during the respective sessions. This is done to make it more interesting as we found deep involvement of the session chairs in mentoring the young scholars during their presentations.

The evolution of ACSS is an interesting process. We have noticed the emergence of security as a very important aspect of research, due to the overwhelming presence of IoT in every aspect of life.

The indexing initiatives from Springer have drawn a large number of high-quality submissions from scholars in India and abroad. ACSS continues with the tradition of the double-blind review process by the PC members and by external reviewers. The reviewers mainly considered the technical aspect and novelty of each paper, besides the validation of each work. This being a doctoral symposium, the clarity of the presentation was also given importance.

The Technical Programme Committee for the symposium selected only 18 papers for publication out of 42 submissions.

We would like to take this opportunity to thank all the members of the Programme Committee and the external reviewers for their excellent and time-bound review works.

We thank the members of the Organizing Committee, whose sincere efforts before and during the symposium have resulted in a friendly and engaging event, where the discussions and suggestions during and after the paper presentations create a sense of community that is so important for supporting the growth of young researchers.

We thank Springer for sponsoring the Best Paper Award. We would also like to thank ACM for the continuous support towards the success of the symposium. We appreciate the initiative and support from Mr. Aninda Bose and his colleagues in Springer Nature for their strong support towards publishing this post-symposium book in the series "Advances in Intelligent Systems and Computing". Last but not least, we thank all the authors without whom the symposium would not have reached up to this standard.

On behalf of the editorial team of ACSS 2019, we sincerely hope that ACSS 2019 and the works discussed in the symposium will be beneficial to all its readers and motivate them towards even better works.

Kolkata, India   Nabendu Chaki
Bialystok, Poland   Khalid Saeed
Kolkata, India   Rituparna Chaki
Venice, Italy   Agostino Cortesi

Contents

About the Editors

Rituparna Chaki is Professor of Information Technology in the University of Calcutta, India. She received her Ph.D. Degree from Jadavpur University in India in 2003. Before this she completed B.Tech. and M.Tech. in Computer Science & Engineering from the University of Calcutta in 1995 and 1997 respectively. She has served as a System Executive in the Ministry of Steel, Government of India for nine years, before joining the academics in 2005 as a Reader of Computer Science & Engineering in the West Bengal University of Technology, India. She is with the University of Calcutta since 2013. Her area of research includes Optical networks, Sensor networks, Mobile ad hoc networks, Internet of Things, Data Mining, etc. She has nearly 100 publications to her credit. Dr. Chaki has also served in the program committees of different international conferences. She has been a regular Visiting Professor in the AGH University of Science & Technology, Poland for the last few years. Dr. Chaki has co-authored a couple of books published by CRC Press, USA.

Agostino Cortesi, Ph.D., is a Full Professor of Computer Science at Ca' Foscari University, Venice, Italy. He served as Dean of the Computer Science studies, as Department Chair, and as Vice-Rector for quality assessment and institutional affairs. His main research interests concern programming languages theory, software engineering, and static analysis techniques, with particular emphasis on security applications. He published more than 110 papers in high-level international journals and proceedings of international conferences. His h-index is 16 according to Scopus, and 24 according to Google Scholar. Agostino served several times as member (or chair) of program committees of international conferences (e.g., SAS, VMCAI, CSF, CISIM, ACM SAC) and he is in the editorial boards of the journals "Computer Languages, Systems and Structures" and "Journal of Universal Computer Science". Currently, he holds the chairs of "Software Engineering", "Program Analysis and Verification", "Computer Networks and Information Systems" and "Data Programming".

Khalid Saeed is a Full Professor in the Faculty of Computer Science, Bialystok University of Technology, Bialystok, Poland. He received the B.Sc. Degree in Electrical and Electronics Engineering in 1976 from Baghdad University in 1976, the M.Sc. and Ph.D. Degrees from Wroclaw University of Technology, in Poland in 1978 and 1981, respectively. He received his D.Sc. Degree (Habilitation) in Computer Science from the Polish Academy of Sciences in Warsaw in 2007. He was a Visiting Professor of Computer Science with the Bialystok University of Technology, where he is now working as a Full Professor He was with AGH University of Science and Technology in the years 2008–2014. He is also working as a Professor with the Faculty of Mathematics and Information Sciences at Warsaw University of Technology. His areas of interest are Biometrics, Image Analysis and Processing and Computer Information Systems. He has published more than 220 publications, edited 28 books, Journals and Conference Proceedings, 10 text and reference books. He supervised more than 130 M.Sc. and 16 Ph.D. theses. He gave more than 40 invited lectures and keynotes in different conferences and universities in Europe, China, India, South Korea and Japan on Biometrics, Image Analysis and Processing. He received more than 20 academic awards. Khalid Saeed is a member of more than 20 editorial boards of international journals and conferences. He is an IEEE Senior Member and has been selected as IEEE Distinguished Speaker for 2011–2016. Khalid Saeed is the Editor-in-Chief of International Journal of Biometrics with Inderscience Publishers.

Nabendu Chaki is a Professor in the Department of Computer Science & Engineering, University of Calcutta, Kolkata, India. Dr. Chaki did his first graduation in Physics from the legendary Presidency College in Kolkata and then in Computer Science & Engineering from the University of Calcutta. He has completed Ph.D. in 2000 from Jadavpur University, India. He is sharing 6 international patents including 4 U.S. patents with his students. Prof. Chaki has been quite active in developing international standards for Software Engineering and Cloud Computing as a member of Global Directory (GD) for ISO-IEC. Besides editing more than 25 book volumes, Nabendu has authored 6 text and research books and has more than 150 Scopus Indexed research papers in Journals and International conferences. His areas of research interests include distributed systems, image processing and software engineering. Dr. Chaki has served as a Research Faculty in the Ph.D. program in Software Engineering in U.S. Naval Postgraduate School, Monterey, CA. He is a visiting faculty member for many Universities in India and abroad. Besides being in the editorial board for several international journals, he has also served in the committees of over 50 international conferences. Prof. Chaki is the founder Chair of ACM Professional Chapter in Kolkata.

Part I
WSN and IoT Applications

Fuzzy Logic-Based Range-Free Localization for Wireless Sensor Networks in Agriculture

Arindam Giri, Subrata Dutta and Sarmistha Neogy

Abstract Knowing the accurate location of nodes in wireless sensor networks (WSN) is a major concern in many location-dependent applications. Especially in agricultural monitoring, accurate location of the affected crop is required for effective use of fertilizers or pesticides. Low-cost and accurate localization algorithms need to be developed for such applications. Range-free localization like DV-Hop has become popular over the years due to its simplicity and that too without complicated hardware. This paper proposes a modified DV-Hop algorithm that improves the accuracy of localization, based on weights of anchors. In this algorithm, fuzzy logic is used to determine the weights of anchors. Simulation results also show that the proposed algorithm improves localization accuracy with respect to DV-Hop algorithm and its variants.

Keywords Sensor network · Localization algorithms · DV-Hop · Fuzzy logic

1 Introduction

Rapid technological advances of low-cost, low-power, and multifunctional sensor devices make them used in every area of life. Generally, deployed wireless sensor networks (WSNs) are used to gather spatiotemporal characteristics of the physical world. They are being successfully used in environmental monitoring, precision agriculture, patient monitoring, habitat monitoring, and object tracking [1]. For instance, WSN used in agriculture [2] can monitor field characteristics like

A. Giri (✉)
Haldia Institute of Technology, Haldia 721657, West Bengal, India
e-mail: ari_giri111@rediffmail.com

S. Dutta
National Institute of Technology, Jamshedpur 831014, Jharkhand, India
e-mail: subrataduttaa@gmail.com

S. Neogy
Jadavpur University, Kolkata 700032, West Bengal, India
e-mail: sarmisthaneogy@gmail.com

© Springer Nature Singapore Pte Ltd. 2020
R. Chaki et al. (eds.), *Advanced Computing and Systems for Security*,
Advances in Intelligent Systems and Computing 995,
https://doi.org/10.1007/978-981-13-8962-7_1

temperature, moisture, nutrients, etc. Location-dependent applications like object tracking need to know the physical location of sensors. Determining the physical position of sensors in WSN is known as localization. Localization plays a major role in such WSN applications. Researchers have contributed much in the field of localization [3]. However, it remains a challenging task due to the limited capacity of tiny sensors like energy and computing power.

Although global positioning system (GPS) provides position information to sensors, it is not feasible to be incorporated in every sensor in terms of cost, hardware, and computational capacity requirements of GPS devices. Again, GPS is not suitable in hazardous environment like industrial plant monitoring as well in indoor applications like habitat monitoring. Generally, in localization algorithms, only a few nodes are enabled with GPS to assess their positions. These nodes are called anchor nodes. The problem of localization is to estimate the locations of unknown nodes in WSN with the help of anchor nodes.

There are many localization algorithms proposed in WSN. The existing localization algorithms can be classified as range-based and range-free. Range-based algorithms estimate locations of sensors based on range measurements like distance [4] or angle [5]. Although range-based localization technique provides accurate location information, they require additional costly hardware for range measurements. Range-free algorithms do not rely on absolute range information, and hence not require additional hardware [6, 7]. Range-free localization algorithms are widely used in large-scale WSN applications like agriculture to avail economic benefit over the other. DV-Hop [8] is a popular range-free localization algorithm. It assumes that anchors closer to an unknown node can provide more accurate hop count than a farther one. However, two anchors closely placed in a randomly distributed WSN consider others in calculating hop count between them. So, a closest anchor not always offers better hop count estimation than far away anchors with accurate hop count. Efforts have been made to DV-Hop algorithm to enhance the accuracy of localization. In weighted DV-Hop, weights of anchors are used to enhance localization accuracy without additional hardware device. The weight of an anchor node is calculated based on minimal hop count between an unknown node and an anchor node. In all DV-Hop variants, weights are assigned to anchors such that the more the hop count, the less is the weight of an anchor. However, this fact is not always true; an anchor farther from an unknown node may give better estimation of hop count possessing more weights than a closely placed anchor. Inaccurate hop count results in inaccurate anchor weights. Hence, deciding weights of anchors with erroneous hop count is a challenge in range-free localization. Uncertainty in deciding weights of anchors can introduce error in localization.

In this paper, we propose a fuzzy-based weighted DV-Hop (called, FWDV-Hop) algorithm where localization accuracy is enhanced dealing with uncertainty in anchor weights. Here, weights are estimated based on erroneous hop count through fuzzy logic [9].

The remainder of this paper is organized as follows. In Sect. 2, DV-Hop algorithm is illustrated along with its variants. We present an overview of the proposed FWDV-Hop algorithm along with an introduction to fuzzy logic in Sect. 3. Simulation of the proposed algorithm and result analysis are presented in Sect. 4, while the conclusion is drawn in Sect. 5.

2 Literature Review

2.1 DV-Hop Algorithm

In this section, DH-Hop algorithm is explained. It is a classical localization algorithm used in WSN which uses a few anchor nodes. The positions of unknown nodes are estimated after getting position information from nearby anchor nodes. DV-Hop is based on the concept of distance vector routing. It works in three phases. In the first phase, all nodes in the network get distances to anchor in hops by exchanging distance vector. In the second phase, each node calculates the average hop size. Then unknown nodes estimate their positions with the help of known anchor positions in the last phase.

In the first phase, each anchor node broadcasts a beacon message providing its position information along with an initial hop count of one. Each receiving node maintains a minimum hop count per anchor for all beacons. Beacons with higher hop count values (called, stale information) from anchors are simply discarded. Non-stale beacons are flooded outward after incrementing hop count at every intermediate hop until all the shortest paths are found. This is how all nodes including the anchors know the shortest path distance to all anchor nodes in terms of minimum hop count information.

In the second phase, each anchor node calculates its average hop size of one hop using the minimum hop count to other anchors. The average hop size of ith anchor is calculated as

$$Hopsize_i = \frac{\sum_{j \neq i} \sqrt{(x_i - x_j)^2 + (y_i - y_j)^2}}{\sum_{j \neq i} h_{ij}} \qquad (1)$$

where (x_i, y_i), (x_j, y_j) are the coordinates of anchors i and j and h_{ij} denotes the minimum hop count between them. After calculating average hop size, an anchor node broadcasts that value throughout the network. Once an unknown node receives hop size, it finds the distance to the beacon node using hop count and hop size using Eq. (2).

$$d_{ij} = Hopsize_i * h_{ij}. \qquad (2)$$

In the last step, an unknown node, P estimates its position, $X = (x, y)$ using position of n number of anchor nodes and the distance, $d_i, i = 1, 2, ..n$, to P as follows:

$$\begin{cases} (x - x_1)^2 + (y - y_1)^2 = d_1^2 \\ (x - x_2)^2 + (y - y_2)^2 = d_2^2 \\ \quad\quad\quad . \quad\quad\quad\quad . \\ (x - x_n)^2 + (y - y_n)^2 = d_n^2 \end{cases} \tag{3}$$

Now, Eq. (3) can be expanded as

$$\begin{cases} 2(x_1 - x_n)x + 2(y_1 - y_n)y = x_1^2 - x_n^2 + y_1^2 - y_n^2 - d_1^2 + d_n^2 \\ 2(x_2 - x_n)x + 2(y_2 - y_n)y = x_2^2 - x_n^2 + y_2^2 - y_n^2 - d_2^2 + d_n^2 \\ \quad\quad\quad\quad\quad\quad . \\ 2(x_{n-1} - x_n)x + 2(y_{n-1} - y_n)y = x_{n-1}^2 - x_n^2 + y_{n-1}^2 - y_n^2 - d_{n-1}^2 + d_n^2 \end{cases} \tag{4}$$

Now, Eq. (4) can be represented by linear system as $AX = B$, where

$$A = -2 \times \begin{bmatrix} x_1 - x_n & y_1 - y_n \\ x_2 - x_n & y_2 - y_n \\ & . \\ & . \\ x_{n-1} - x_n & y_{n-1} - y_n \end{bmatrix},$$

$$B = \begin{bmatrix} d_1^2 - d_n^2 - x_1^2 + x_n^2 - y_1^2 + y_n^2 \\ d_2^2 - d_n^2 - x_2^2 + x_n^2 - y_2^2 + y_n^2 \\ . \\ . \\ d_{n-1}^2 - d_n^2 - x_{n-1}^2 + x_n^2 - y_{n-1}^2 + y_n^2 \end{bmatrix}, \text{ and}$$

$$X = \begin{bmatrix} x \\ y \end{bmatrix}$$

Hence, X can be calculated as

$$X = \left(A^T A\right)^{-1} A^T B. \tag{5}$$

The limitation of DV-Hop algorithm is that it assumes that the anchor node closest to unknown node can contribute accurate hop size estimation than others. However, in a randomly deployed WSN, anchor nodes situated very close to each other add up nearby anchors in their hop count calculation. In this situation, an anchor closest to a sensor node does not necessarily give accurate hop size estimation. Instead, an anchor far away from an unknown node with appropriate hop size may offer better estimation of coordinates. Hence, location estimation of WSN nodes is affected by this approach. To get rid of the problem, the anchors are assigned weights reflecting their impact on location estimation for an unknown node.

2.2 Variants of DV-Hop Algorithm

In weighted DV-Hop algorithm [10], the weight of an anchor is calculated based on the hop count as below:

$$w_i = \frac{1}{h_{ij}}. \tag{6}$$

Here, w_i's might not be accurately calculated if hop count is inaccurate. Unlike DV-Hop, the weighted DV-Hop algorithm uses weights of anchors in a least square method in Eq. (7) as

$$f(x, y) = min \sum_{i=1}^{n} w_i^2 (\sqrt{(x_i - x_j)^2 + (y_i - y_j)^2} - d_i)^2, \tag{7}$$

where w_i is the weight of anchor node i and n is number of nodes. In weighted DV-Hop, the value of w_i is set using Eq. (6). Now, Eq. (7) can be expressed as $A'X = B'$, where

$$A' = -2 \times \begin{bmatrix} w_1^2 w_n^2 (x_1 - x_n) & w_1^2 w_n^2 (y_1 - y_n)_1 \\ w_2^2 w_n^2 (x_2 - x_n) & w_2^2 w_n^2 (y_2 - y_n) \\ & \cdot \\ & \cdot \\ & \cdot \\ w_{n-1}^2 w_n^2 (x_{n-1} - x_n) & w_{n-1}^2 w_n^2 (y_{n-1} - y_n) \end{bmatrix}, \text{ and}$$

$$B' = \begin{bmatrix} w_1^2 w_n^2 (d_1^2 - d_n^2 - x_1^2 + x_n^2 - y_1^2 + y_n^2) \\ w_2^2 w_n^2 (d_2^2 - d_n^2 - x_2^2 + x_n^2 - y_2^2 + y_n^2) \\ \cdot \\ \cdot \\ w_{n-1}^2 w_n^2 (d_{n-1}^2 - d_n^2 - x_{n-1}^2 + x_n^2 - y_{n-1}^2 + y_n^2) \end{bmatrix}$$

Then, the position of unknown node is estimated as

$$X = \left(A'^T A' \right)^{-1} A'^T B'. \tag{8}$$

Improved weighted DV-Hop (IWDV-Hop), proposed by Guadane et al. [10], is an improved version of DV-Hop algorithm which uses inversely weighted hop count in location estimation. It considers not only a closer anchor to estimate unknown location but also it uses a far away anchor with accurate hop count.

$$w_i = \frac{\frac{1}{h_{ij}}}{\sum_{k=1}^{n} \frac{1}{h_{kj}}}, k \neq j. \tag{9}$$

The root mean square error (RMSE) in location estimation of a node i is calculated in Eq. (10), assuming (x, y) and (x_i, y_i) as the actual and estimated coordinates of i, respectively.

$$e = \frac{1}{n-1} \sum_{i=1}^{n} \sqrt{(x - x_i)^2 + (y - y_i)^2}. \tag{10}$$

In [11], each anchor node calculates its own hop size as the weighted sum of all average one-hop distances (HWDV-Hop) to all anchors. The proposed algorithm estimates the average hop size for the network by average hop size weighted mean. HWDV-Hop demands that it outperforms DV-Hop in terms of localization accuracy.

3 Overview of FWDV-Hop Algorithm

This section gives an overview of fuzzy-based weighted DV-Hop (FWDV-Hop) algorithm. In DV-Hop, an unknown node estimates its location using least square method as in Eq. (3). Weights are assigned to anchors to impose the relative importance over other anchors. Weights calculated directly based on inaccurate/noisy hop count provides inaccurate average hop size which decreases the accuracy of localization. In the proposed FWDV-Hop, we assign weights of anchors using fuzzy logic. This approach can provide appropriate weight assignment so as to localize nodes dealing with uncertainty in hop count. This algorithm follows DV-Hop except the weight calculation for anchors through fuzzy inference system. The fuzzy input hop count is feed to FIS in Fig. 1 and mapped to weight via inference engine using fuzzy rules. The membership function of fuzzy variables hop count and weight is divided into five triangular membership functions such as very low (VL), low (L), moderate (M), high (H), and very high (VH) as shown in Fig. 2.

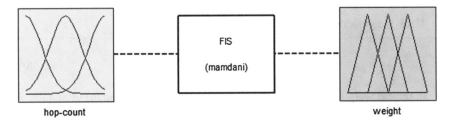

Fig. 1 Fuzzy model for proposed algorithm with hop count as input and weight as output

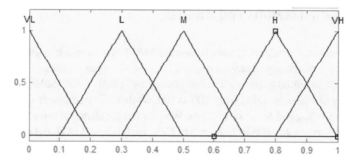

Fig. 2 Membership functions for input/output variable

3.1 Introduction to Fuzzy Logic

Fuzzy logic [9], proposed by Zadeh, is mainly used in linguistic reasoning. It resembles the logic of human thought in a flexible computation than computers. It demands to be used in application environments dealing with imprecise, vague, and erroneous information. Fuzzy logic can better represent the vague data like "high" with fuzzy set and fuzzy membership function. Fuzzy sets can prescribe a range of values, called domain characterized by the membership function. A fuzzy logic system, called fuzzy inference system (FIS), consists of three components: fuzzifier, inference engine, and defuzzifier. A fuzzifier maps each crisp input to a fuzzy set and assigns a value equal to the degree of membership. Inference engine is built upon IF-ELSE rules and inference methods. Given a number of fuzzy inputs, the inference engine calculates fuzzy output based on fired rule. Finally, the defuzzifier maps the fuzzy output to crisp output. The fuzzy inference system is depicted in Fig. 3.

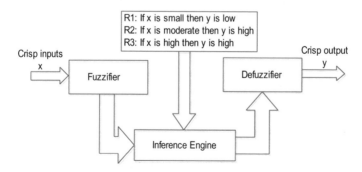

Fig. 3 A fuzzy inference system showing three components

4   Simulation Results and Analysis

For implementing our algorithm, we have used MATLAB. A wireless sensor network is created with randomly deployed sensors and anchors over an area of 50×50 m^2. All nodes are assumed to have 2 J initial energy and 15 m of communication radius. The number of nodes is varied from 100 to 500, while radio communication radius of each sensor is changed from 10 to 20 m. We plot the localization error with varying anchor nodes to sensor node ratio from 5 to 40%. In the same MATLAB environment, other algorithms are evaluated. Performance of the proposed algorithm is measured through normalized localization error (RMSE) using Eq. (10) as the average deviation of estimated location to actual location of all unknown nodes. The simulation results are used to plot localization error varying number of nodes, communication radius, and ration of anchor nodes. Proposed algorithm is compared with other algorithms in Fig. 4 through Fig. 6.

Figure 4 depicts localization error with respect to total number of sensor nodes, with 10% of anchor nodes. It is observed that localization error decreases as number of increases. The error becomes steady when the number of sensors increases to 300 and more. With the same number of nodes and 10% of anchors, the proposed algorithm outperforms others as the weights are fixed with fuzzy logic.

In Fig. 5, localization error is plotted with varying anchor nodes from 5 to 40% of total nodes in the network. The more is the anchor ratio, the less is the localization error. Dealing uncertainties through fuzzy logic, our algorithm gives better localization accuracy than others.

In Fig. 6, localization error is depicted with respect to communication radius of sensors varying from 10 to 20 m. Here, 100 nodes are randomly deployed with 20% of anchor nodes in order to evaluate algorithms. With increasing radius of sensors, localization accuracy is improved. Again, proposed algorithm outperforms others in terms of localization accuracy due to appropriate fixing of anchor weights in location estimation.

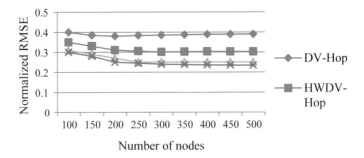

Fig. 4 Normalized localization error versus number of sensor nodes, with 10% of anchor nodes

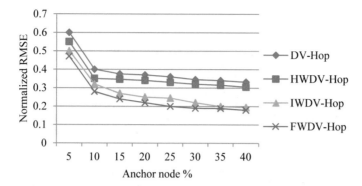

Fig. 5 Normalized localization error versus anchor nodes ratio (in percentage of anchor nodes) with 100 sensor nodes

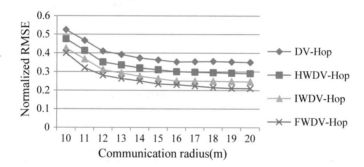

Fig. 6 Normalized localization error versus radio range of sensors, with 100 sensors and 10% of anchor nodes

5   Conclusion

Localization is essential to location-dependent applications like agriculture monitoring. Range-free localization becomes popular because of its no additional hardware requirements. In this paper, a range-free localization method is presented using fuzzy logic. In weighted DV-Hop algorithm, weights are calculated as the inverse of hop count. With closely placed anchor nodes in randomly distributed sensor network, hop count estimation is not correct, so weights are affected. In this research, fuzzy logic is utilized to find weights of anchors based on erroneous hop count. Localization error is calculated for algorithms with varying number of sensors, anchors, and communication radius of sensors. Simulation results reveal that the proposed algorithm improves others in terms of localization accuracy.

References

1. Akyildiz, I.F., Su, W., Sankarasubramaniam, Y., Cayirci, E.: Wireless sensor networks: a survey. Comput. Netw. **38**(4), 393–422 (2002)
2. Giri, A., Dutta, S., Neogy, S., Dahal, K., Pervez, Z.: Internet of things (IoT): a survey on architecture, enabling technologies, applications and challenges. In: Proceedings of the 1st International Conference on Internet of Things and Machine Learning, pp. 7:1–7:12 (2017)
3. Patwari, N., Ash, J.N., Kyperountas, S., Hero, A.O., Moses, R.L., Correal, N.S.: Locating the nodes: cooperative localization in wireless sensor networks. IEEE Signal Process. Mag. **22**(4), 54–69 (2005)
4. Shanshan, W., Jianping, Y., Zhiping, C., Guomin, Z.: A RSSI-based self-localization algorithm for wireless sensor networks. J. Comput. Res. Dev. **45**(1), 385–388 (2008)
5. Niculescu, D., Nath, B.: Ad hoc positioning system (APS) using AOA. In: INFOCOM 2003. Twenty-Second Annual Joint Conference of the IEEE Computer and Communications. IEEE Societies, vol. 3, pp. 1734–1743 (2003)
6. Shang, Y., Ruml, W.: Improved MDS-based localization. IEEE Infocom **4**, 2640–2651 (2004)
7. He, T., Huang, C., Blum, B.M., Stankovic, J.A., Abdelzaher, T.: Range-free localization schemes for large scale sensor networks. In: Proceedings of the 9th annual international conference on Mobile computing and networking, pp. 81–95 (2003)
8. Niculescu, D., Nath, B.: Ad hoc positioning system (APS). In: Global Telecommunications Conference, 2001. GLOBECOM'01. IEEE, vol. 5, pp. 2926–2931 (2001)
9. Zadeh, L.A.: Fuzzy logic = computing with words. IEEE Trans. Fuzzy Syst. **4**(2), 103–111 (1996)
10. Guadane, M., Bchimi, W., Samet, A., Affes, S.: Enhanced range-free localization in wireless sensor networks using a new weighted hop-size estimation technique. In: 2017 IEEE 28th Annual International Symposium on Personal, Indoor, and Mobile Radio Communications (PIMRC), pp. 1–5 (2017)
11. Hadir, A., Zine-Dine, K., Bakhouya, M., El Kafi, J.: An optimized DV-hop localization algorithm using average hop weighted mean in WSNs. In: 2014 5th Workshop on Codes, Cryptography and Communication Systems (WCCCS), pp. 25–29 (2014)

End-User Position-Driven Small Base Station Placement for Indoor Communication

Anindita Kundu⊙ , Shaunak Mukherjee, Ashmi Banerjee
and Subhashis Majumder⊙

Abstract With the increase in the number of wireless end users, the demand for high-speed wireless network has increased by multiple folds. Moreover, most of the data traffic is observed to be generated from the indoor environment. Hence, researchers have come up with the solution of deploying Small Base Stations (SBSs) in the indoor environment, which proves to be immensely effective in providing the last mile connectivity. However, since the deployment of the SBSs is unplanned, there exists a high chance of co-tier interference, which might be mitigated by transmission power control but at the cost of degraded quality of service for some of the end users. Hence, in this work, we propose the concept of mobile SBSs which are connected to power grids located in the ceiling of the deployment region. The mobile SBSs position themselves at the received signal strength-based centroids of the end user clusters, thereby mitigating the co-tier interference as well as conserving power of the handsets by minimizing the distance between the SBS and the end-user handsets. To enhance the security of the system, the SBSs are considered to form a closed group. The proposed system requires about 13% lesser number of SBSs while enhancing the end-user coverage by 9.3%. Moreover, about 10% improvement has been observed in terms of cumulative throughput for mobile SBSs compared to fixed SBSs. Thus, the deployment of mobile SBSs proves to be more effective compared to the fixed SBSs for indoor communication.

A. Kundu (✉) · S. Majumder
Heritage Institute of Technology, Kolkata, India
e-mail: anindita.kundu@heritageit.edu

S. Majumder
e-mail: subhashis.majumder@heritageit.edu

S. Mukherjee
Whiting School of Engineering, Johns Hopkins University, Baltimore, USA
e-mail: shaunak.mukherjee94@gmail.com

A. Banerjee
Technische Universität München, Munich, Germany
e-mail: ashmi.banerjee@tum.de

© Springer Nature Singapore Pte Ltd. 2020
R. Chaki et al. (eds.), *Advanced Computing and Systems for Security*,
Advances in Intelligent Systems and Computing 995,
https://doi.org/10.1007/978-981-13-8962-7_2

Keywords 5G network · Small cells · Mobile base station placement · K-means clustering algorithm · Shortest path avoiding obstacles

1 Introduction

As the number of wireless end users are increasing, the demand for high-speed communication is also increasing. However, recent research exhibits that most of the data traffic is usually generated in the indoor environment [18]. 5G assures to satisfy this demand and provide a high-speed communication platform using millimeter waves, small cells, MIMO, etc. [21, 23]. Small cells have gained immense popularity in recent years in providing satisfactory network coverage in the indoor environment [13, 18]. They also participate in energy conservation by transmitting at a very low Transmission Power (TP) [5].

Usually, the small cells are considered to be deployed in the end-user premises by the end-user themselves and mostly in an unplanned fashion [10, 24]. Thus, unknowingly two families residing in two adjacent apartments may place their Small Base Stations (SBSs) on either side of the same wall which will lead to significant amount of co-tier interference between the two, leading to a much degraded performance. Also, the fixed, unplanned deployment of the small cells may lead to scenarios where the end users may be located in a cluster at the cell edge which in turn will demand the highest TP of the serving SBS. The transmission of the SBSs at their highest TP may again lead to enhanced co-tier interference. Thus, as a solution to this problem, we propose the deployment of mobile SBSs. This solution is applicable to both residential and commercial spaces. To provide enhanced security, the SBSs are considered to form a *closed group* such that they can serve only verified end users [7].

A power grid is assumed to exist on the ceiling of the buildings where the coverage is to be provided. It is considered that the SBSs can be mounted on the grid and are also movable along the grid. The initial positions of deployment of the SBSs are identified using the PSOM algorithm [14]. On identifying the positions, the SBSs are mounted on the power grid at the positions identified by the PSOM algorithm. The SBSs within a specific deployment region are all considered to be connected to a special SBS which also acts as the Central Intelligent Controller (CIC). The CIC may be located anywhere in the deployment region and is connected to the rest of the SBSs via the wired power grid located at the ceiling. It monitors the position of the other SBSs as well as the active end users via the SBSs. Initially, the SBSs are all in low-powered *sleep* mode but monitor the end-user requests. Whenever an end user requests for service, while the nearest SBSs is in sleep mode, the SBS counts the number of times the request is being sent from the end user (Request Count-RC). If the RC is below the desired threshold (RC_{Th}) of 15 [15], the SBS continues in the *sleep* mode assuming that it may be some other SBS which is in the *active* state and will take care of the end user. However, once $RC = RC_{Th}$, the SBS turns itself *on* and passes on the information to the CIC. The CIC then considers the cluster of the active end users and identifies the Cluster Center (CC) using the RSS-based K-means clustering algorithm [1].

On finding the CC, the CIC identifies an A* algorithm-based shortest path, while avoiding the obstacles [8] to identify the nearest SBS to the CC. It then instructs the concerned SBS to follow the obstacle-free shortest path identified and place itself at the desired CC. To evaluate the performance of the network after the deployment of the SBS, the coverage and throughput of the system are next calculated. If either coverage or throughput is observed to be below the desired threshold, the CIC partitions the end-user cluster into further subclusters and thus identifies new CCs. The SBSs located nearest (in terms of obstacle-free path) to these new CCs are then instructed to move to CCs. More SBSs are activated if required. This process continues iteratively until all the end users have been accommodated or as long as the number of available SBSs do not get exhausted. On positioning the SBSs, a distributed TP control algorithm [15] is used by the SBSs to identify the minimum TP required to satisfy the end users.

Results exhibit that mobile SBSs require about 13% lesser number of SBSs while enhancing the end users' coverage by 9.3% compared to the fixed SBSs. Moreover, about 10% improvement has been observed in terms of cumulative throughput for mobile SBSs. Thus, the mobile SBSs prove to be more effective compared to fixed SBSs for indoor communication.

The rest of this paper is organized as follows. Section 2 discusses the existing works in the literature, while Sect. 3 describes the scenario considered. Section 4 illustrates the methodology adopted. The proposed algorithm is illustrated in Sect. 5 followed by the discussion of the results obtained in Sect. 6. Finally, Sect. 7 concludes the work.

2 Related Work

Small cells have been of great interest to the researchers for quite some time now [2, 9, 11, 13, 22]. Ultradense networks comprising of small cells co-existing with traditional macro-cells is one of the major features of the evolving 5G networks. In spite of the various advantages of using small cells in terms of capacity enhancement, cost-effectiveness, and energy efficiency, there are some major challenges that need to be addressed. Maintaining an economic and ubiquitous connectivity among the macro-cells and the small cells is one of them. Other significant challenges include co-tier and cross-tier interference, energy efficiency, reliability, and maintenance of the SBSs and security.

In this work, we have taken up the problem of placement of SBSs. The proposed system ensures that the co-tier and cross-tier interference is minimized while improving the energy efficiency of the system. Similar works have been taken up by researchers in recent literature [3, 12, 20]. A Mixed Integer Nonlinear Programing (MINLP) approach has been taken by authors of [12] to understand the various interference and multiplexing patterns at sub-6 GHz and microwave band. It is then solved using linear relaxation and branch-and-bound algorithm and is applied in a wireless backhaul network. The experiment conducted by the authors of [25] exhibits that densely deployed dedicated indoor femtocells are more efficient in terms of spectrum and energy and provides a solution to the huge indoor capacity demands. Hence, we have taken up the problem of placement of SBSs.

However, all of the aforementioned algorithms have considered the deployment of fixed SBSs. Fixed SBSs may be sufficient for the traffic load at the time of its deployment but may not satisfy the increasing demand with chores of time. It may also prove to be inefficient in cases where the end users form nonuniform clusters. Hence, we have proposed the concept of mobile SBSs which places itself at the centroid of the clusters formed by the end users. The movement of the SBSs is controlled by the CIC. There are various algorithms in the literature that deal with finding the shortest path between two specific locations while avoiding the obstacles between the nodes [4, 17]. However, traditional breadth-first search, Dijkstra's algorithm, and A* algorithms also provide the desired solutions [6]. Each of these algorithms has its pros and cons. Since we are dealing with a graph almost similar to a grid, we consider the A* algorithm to identify the shortest obstacle-free path between the old and new positions of the SBSs.

3   Scenario Description

To model the scenario, the floor plan is system generated based on a random binary matrix called **GRID** whose dimensions indicate the plan of the campus. Each cell of the matrix represents the presence/absence of an obstacle in a unit area of the deployment region considered. If a cell has value 1, it indicates the presence of an obstacle, while a 0 indicates a free area. A cell of the matrix corresponds to 1 m × 1 m region of the deployment area. N number of SBSs are initially considered to be placed in a random fashion in the deployment region. Initially, these SBSs are considered to be in the *sleep* mode. Now, n number of active calls are considered to be generated from the randomly placed end users within the system. Figure 1 illustrates such a randomly generated scenario.

4   Methodology Adopted

4.1   Forming the Graph of the Deployment Region

For the proposed algorithm to work in the CIC, the first job that needs to be done is to model the deployment region. In order to model the deployment region, first a random binary matrix GRID is generated. Each cell of the GRID corresponds to 1 sq. m. area of the actual deployment region. If the entry in the cell is 1, it implies

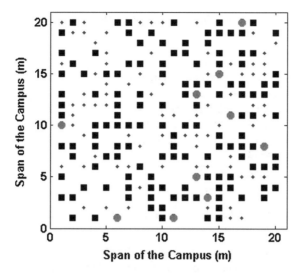

Fig. 1 Initial scenario of the deployment region where the black squares indicate the obstacles, the blue dots indicate the end users, while the big magenta dots indicate the SBSs

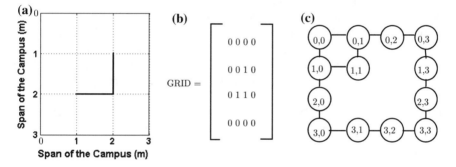

Fig. 2 Modeling the deployment region. **a** Plan of the deployment region. **b** The GRID matrix corresponding to the plan of the deployment region. **c** Undirected graph corresponding to the plan of the deployment region

that an obstacle is present in the actual deployment region corresponding to it while a 0 implies free space.

For example, Fig. 2b exhibits the content of the GRID matrix corresponding to a sample region given in Fig. 2a. Next, in order to find the shortest obstacle-free path, the layout of the deployment region is converted into an undirected graph as shown in Fig. 2c. In order to proceed with it, each section of the deployment region is represented by a node, and two nodes are connected by an edge if their corresponding sections are vertically or horizontally adjacent.

4.2 Identifying the Initial Positions of the SBSs and the End Users

After the deployment region has been modeled, the N SBSs are deployed randomly. Initially, all of them are considered to be in the *sleep* mode. Their Cartesian coordinate positions with respect to the **GRID** matrix are updated in the CIC in the $N \times 2$ matrix called **SBS_POS**. Similarly, the end users are deployed randomly and their Cartesian coordinate positions with respect to the **GRID** matrix are stored in the $n \times 2$ matrix called the **USER_POS**.

4.3 Identifying the Cluster Centers (CC) Positions

Once the position of the end users has been identified, the CIC employs a modified version of the K-means clustering algorithm to divide them into K clusters based on their Received Signal Strength (RSS) [1]. The K-value initially starts with 2^0 and increases with every iteration. The coverage of the end users (C) and their cumulative throughput (T) are calculated at every iteration. If either C or T fails to achieve the desired threshold, then K is incremented exponentially (with increasing powers of 2) and the RSS-based K-means clustering algorithm is executed to identify the next set of CCs. The process continues till both the constraints of C and T are satisfied. Once the satisfaction level is achieved for C and T, a binary search like method is executed within the K-values ranging from $K/2$ to K to identify the minimum K-value for which the constraints of C and T are satisfied. The CCs thus identified form the set of desired SBS positions for the iteration. The CIC then instructs the K-nearest SBSs (in terms of obstacle-free shortest path) to move to the CCs.

4.4 Calculating the RSS

The RSS of each end user is a function of their distance (d) from the CC and is calculated using Eq. (1) where the expression for path loss is given by Eq. (2) [16].

$$RSS[dB] = TP[dB] - PL(d)[dB] \tag{1}$$

$$PL(d)[dB] = PL(d_0) + 10\beta.log_{10}(\frac{d}{d_0}) + X_\sigma \tag{2}$$

In Eq. (2), d_0 indicates the free space reference distance of 1 m, while β indicates the optimized pathloss exponent (PLE) for a particular frequency band or environment and X_σ indicates a zero mean Gaussian random variable with standard deviation σ in dB. The height of the SBSs is considered to remain fixed at 7 m, and their operating center frequency is considered to be 28.0 GHz.

4.5   Calculating the Coverage (C)

An end user is considered to be covered only if its RSS requirement, interference requirement as well as its channel requirement are satisfied. Considering that K SBSs are required to be deployed out of which j is the serving SBS, the RSS of each end user is calculated using Eq. (1). The interference is considered to be the cumulative sum of the RSS from the $K - 1$ neighboring non-serving SBSs and is given in Eq. (3).

$$Interference[dB] = \sum_{i=1,i\neq j}^{K} RSS_i \qquad (3)$$

Each SBS is considered to have a fixed number of channels (c) of equal bandwidth (b). If the cumulative channel requirement of a group of end users to be served by an SBS is more than c, then the end users whose channel requirement could not be satisfied is considered to be uncovered in spite of their RSS and interference constraint being satisfied. In such cases, the RSS-based K-means clustering algorithm executes again with increased K-value until the desired constraints are satisfied. Thus, the coverage of the system is calculated as the percentage of end users whose RSS, interference as well as the channel constraints are satisfied.

4.6   Calculating the Throughput (T)

T of the system indicates the cumulative throughput achieved by the end users in the system. The individual throughput of the end users is calculated using Shannon–Hartley theorem [19]. Thus, the T of the system is as given by Eq. (4) where B indicates the bandwidth, SNR indicates the signal-to-noise ratio where noise refers to Additive White Gaussian Noise (AWGN).

$$T = \sum_{i=1}^{n} (B log_2(1 + SNR))_i \qquad (4)$$

4.7   Identifying the Obstacle-Free Shortest Path

The obstacle-free shortest path from the actual position of an existing SBS to its new position identified by the CIC is calculated by traversing the graph given in Fig. 2c using the A* algorithm. For each active SBS, if the current Cartesian position is given by the coordinates (x, y), the node (x, y) in the graph (refer Fig. 2c) is considered to be the source node. If the destination Cartesian position is (p, q), then the node (p, q) in the graph is identified by the CIC as the destination node. Any intermediate node

that lies between the source node and the destination node in the graph is identified as
(x', y'), and the expected cost to the destination value of the node (x', y') is considered
to be the Manhattan distance between the nodes (x', y') and (p, q).

5 Proposed Algorithm

The proposed algorithm has been designed based on the methodology discussed in
the previous section and is given in Algorithm 1. It is considered to be executed
in the CIC with the help of the SBSs to identify their desired positions within the
deployment region based on the end-user positions.

Algorithm 1 Proposed Algorithm to place the Mobile SBSs

1: Form the **GRID** matrix and the connection graph of the deployment region
2: Identify the positions of the randomly deployed SBSs and update the **SBS_POS** matrix
3: Update the identity of each active SBS in the list of $Active\_SBSs$
4: Identify the positions of the active end users and update the **USER_POS** matrix
5: Set $K \leftarrow 1$
6: **while** true **do**
7:    Partition the end users into K clusters
8:    Calculate the RSS of the end users
9:    Calculate the $Interference$ of the end users
10:    Calculate the C of the system
11:    Calculate the T of the system
12:    **if** $C \geq C_{Th}$ and $T \geq T_{Th}$ **then**
13:        Break
14:    **else**
15:        $K \leftarrow K \times 2$
16:    **end if**
17: **end while**
18: K_{opt}=FIND_K_Bin_Search($K/2$, K)
19: **Desired Pos** = Kmeans($GRID$, **SBS_POS**, **USER_POS**, K_{opt})
20: **PATH**=find_shortest_path($Active\_SBSs$, **SBS_POS**, **Desired Pos**)
21: Move the active SBSs to their desired new positions
22: Update **SBS_POS**

---

6 Results and Discussion

The aforementioned scenario has been simulated exhaustively in MATLAB R2014a
with a typical set of simulation parameters as given in Table 1. Further, the efficacy of
the proposed algorithm has been evaluated in terms of the number of SBSs required
to be deployed in order to satisfy the demand of the end users, the coverage provided
by the SBSs and throughput of the system. Figures 3, 4, 5 and 6 illustrate the results.

Table 1 Simulation parameters

Parameter	Value
C_{Th}	95%
T_{Th}	50% of available bandwidth
Span of the Campus	$20\,\text{m} \times 20\,\text{m}$
No. of SBSs deployed in the scenario	10
RSS_{Th}	$-100\,\text{dBm}$
$Interference_{Th}$	$-100\,\text{dBm}$
Initial TP of the SBSs	$10\,\text{dBm}$
No. of channels available at each SBS	8

Fig. 3 Number of SBSs required for varied number of active end users

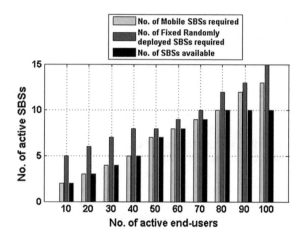

In the aforementioned scenario, initially, the end users are considered to be deployed in a completely random fashion throughout the deployment region. The requirement of the number of active fixed SBSs to serve the end users is much higher compared to the requirement of active mobile SBSs (shown in Fig. 3). Since the CIC partitions the end users into clusters depending upon their RSS and places the mobile SBSs exactly at the centroid positions, it assures that the calls of maximum number of end users are accommodated. This behavior remains unchanged for varied end-user density scenarios. However, in higher end-user density scenarios, it is observed that the requirement of the number of mobile SBS is much higher than available. This is because each SBS is considered to have at most 8 number of channels which are shared among the end users if the number of active end users being served by the SBS is less than or equal to 8. In case, if the number exceeds 8, the SBS fails to accommodate the 9th call leading to a call block, which in turn affects the system capacity. Since 10 mobile SBSs are considered to be deployed in the scenario, the CIC tries to accommodate all the end users using these 10 mobile SBSs. However, as and when the number of end users exceeds $8 \times 10 = 80$, the system fails to accommodate the calls.

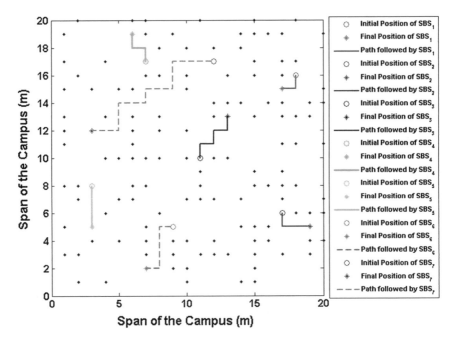

Fig. 4 Path identified by the A* algorithm for the 7 SBSs required as the number of active end users increases from 40 to 50

As observed in Fig. 3, increase in the number of end users triggers the requirement of increased number of active SBSs. For a scenario with 40 end users, the required number of SBSs was 5. However, as the number of active end users increases from 40 to 50, i.e., 10 new end users' request service, 2 more SBSs are required to provide service. The CIC thus activates two SBSs who were in the sleep mode and reorients the SBSs based on their last known stable position and the end-user positions. Figure 4 illustrates the path undertaken by the five active and the two newly activated SBSs such that the cumulative distance traversed by them is minimized. The paths followed by the already active SBSs are marked by solid lines, while the paths followed by the newly activated SBSs are marked by dotted lines in the figure. Moreover, the o symbol marks the initial position of an SBS, i.e., the position w.r.t 40 end users and the * symbol marks the final position of an SBS for the specific positions of the end users, i.e., the position w.r.t. 50 users.

The coverage of the system, as shown in Fig. 5 depends upon the RSS, interference, and the number of channels granted to the end users. In case of lower end-user density scenarios, it is observed that all the end users can be served. However, in medium to higher user density scenarios, the fixed SBSs are observed to behave poorly compared to the mobile SBSs. There are two reasons primarily for this behavior. First, as the number of the fixed SBSs increases, the interference experienced by the end users increase drastically leading to the failure of their interference constraint. However, in case of the mobile SBSs, since they have the capacity for spatial movement,

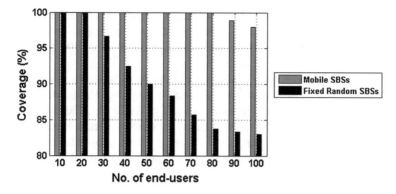

Fig. 5 Coverage provided to the end users

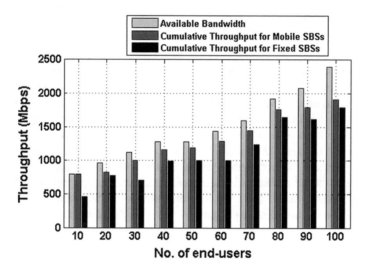

Fig. 6 Cumulative throughput of the network

the interference constraint of the end users in the scenario considered does not fail. Moreover, since each SBS can accommodate only eight end users, in higher end-user density scenarios where the number of requesting end users exceeds the available number of channels, call blocking occurs leading to the drop in coverage in case of both types of SBSs.

The throughput of the system as shown in Fig. 6 is a function of the available bandwidth, number of active end users, and their SNR. As the user density increases, the number of active end users may also increase. At the same time, the required number of SBSs also increases, which in turn increases the available bandwidth. Hence, the throughput of the system increases with the end-user density. However, for higher user density scenarios, the available bandwidth falls compared to the number of active end users resulting in lowering of the rate of increase of the throughput. The throughput observed for the fixed SBSs is much lower compared to that obtained for mobile SBSs as the RSS for the end users in case of fixed SBS is much lower compared to the RSS in case of mobile SBSs.

In case of sparsely populated scenario, for example, 10 users, the number of active SBSs required is 2 as exhibited in Fig. 3. Thus in case the number of users in less than 10, depending upon their positions either 1 or 2 SBSs will be active. In such scenarios as observed from Fig. 5, we can say that their coverage is expected to be 100%. Moreover, similar behavior is also observed in case of cumulative throughput. As exhibited in Fig. 6, their cumulative throughput will be equal to the available bandwidth of the network.

7 Conclusion

Recent research reports that most of the data traffic is generated from the indoor environment and yet indoor communication suffers from various problems like dead zones and channel shortage. To enhance the indoor coverage, in this work, we propose a mobile small base station placement algorithm. The performance of the proposed algorithm has been compared to a system comprising of randomly deployed fixed SBSs and has been observed to outperform it. About 13% improvement has been noted in terms of the number of SBSs required, while about 9.3% improvement has been observed in terms of coverage provided to the end users. Moreover, about 10% improvement has been observed in terms of cumulative throughput for mobile SBSs compared to fixed SBSs. Thus, we can conclude that deployment of mobile SBSs will prove to be more effective compared to fixed SBSs for indoor communication.

References

1. Altintas, B., Serif, T.: Improving rss-based indoor positioning algorithm via k-means clustering. In: Wireless Conference 2011-Sustainable Wireless Technologies (European Wireless), 11th European, pp. 1–5. VDE (2011)
2. Bartoli, G., Fantacci, R., Letaief, K., Marabissi, D., Privitera, N., Pucci, M., Zhang, J.: Beamforming for small cell deployment in lte-advanced and beyond. IEEE Wirel. Commun. 21(2), 50–56 (2014)
3. Bhushan, N., Li, J., Malladi, D., Gilmore, R., Brenner, D., Damnjanovic, A., Sukhavasi, R., Patel, C., Geirhofer, S.: Network densification: the dominant theme for wireless evolution into 5g. IEEE Commun. Mag. 52(2), 82–89 (2014)
4. Brass, P., Vigan, I., Xu, N.: Shortest path planning for a tethered robot. Comput. Geom. 48(9), 732–742 (2015)
5. Bu, S., Yu, F.R.: Green cognitive mobile networks with small cells for multimedia communications in the smart grid environment. IEEE Trans. Veh. Technol. 63(5), 2115–2126 (2014)
6. Cormen, T.H., Leiserson, C.E., Rivest, R.L., Stein, C.: Introduction to Algorithms. MIT press (2009)

7. De La Roche, G., Valcarce, A., López-Pérez, D., Zhang, J.: Access control mechanisms for femtocells. IEEE Commun. Mag. **48**(1) (2010)
8. Duchoň, F., Babinec, A., Kajan, M., Beňo, P., Florek, M., Fico, T., Jurišica, L.: Path planning with modified a star algorithm for a mobile robot. Procedia Eng. **96**, 59–69 (2014)
9. ElSawy, H., Hossain, E., Kim, D.I.: Hetnets with cognitive small cells: user offloading and distributed channel access techniques. IEEE Commun. Mag. **51**(6), 28–36 (2013)
10. Ge, X., Tu, S., Mao, G., Wang, C.X., Han, T.: 5g ultra-dense cellular networks. IEEE Wirel. Commun. **23**(1), 72–79 (2016)
11. Hoadley, J., Maveddat, P.: Enabling small cell deployment with hetnet. IEEE Wirel. Commun. **19**(2), 4–5 (2012)
12. Islam, M.N., Sampath, A., Maharshi, A., Koymen, O., Mandayam, N.B.: Wireless backhaul node placement for small cell networks. In: 2014 48th Annual Conference on Information Sciences and Systems (CISS), pp. 1–6. IEEE (2014)
13. Jungnickel, V., Manolakis, K., Zirwas, W., Panzner, B., Braun, V., Lossow, M., Sternad, M., Apelfrojd, R., Svensson, T.: The role of small cells, coordinated multipoint, and massive mimo in 5g. IEEE Commun. Mag. **52**(5), 44–51 (2014)
14. Kundu, A., Majumder, S., Misra, I.S., Sanyal, S.K.: A heuristic approach towards minimizing resource allocation for femto base station deployment. In: 2016 European Conference on Networks and Communications (EuCNC), pp. 330–334. IEEE (2016)
15. Kundu, A., Majumder, S., Misra, I.S., Sanyal, S.K.: Distributed heuristic adaptive power control algorithms in femto cellular networks for improved performance. Trans. Emerg. Telecommun. Technol. **29**(5), e3280 (2018)
16. MacCartney, G.R., Rappaport, T.S., Samimi, M.K., Sun, S.: Millimeter-wave omnidirectional path loss data for small cell 5g channel modeling. IEEE Access **3**, 1573–1580 (2015)
17. Magadevi, N., Kumar, V.J.S.: Energy efficient, obstacle avoidance path planning trajectory for localization in wireless sensor network. Cluster Comput., 1–7 (2017)
18. Maiti, P., Mitra, D.: Explore tv white space for indoor small cells deployment and practical pathloss measurement. In: 2017 International Conference on Innovations in Electronics, Signal Processing and Communication (IESC), pp. 79–84. IEEE (2017)
19. Rappaport, T.S., et al.: Wireless Communications: Principles and Practice, vol. 2. prentice hall PTR New Jersey (1996)
20. Richter, F., Fettweis, G.: Cellular mobile network densification utilizing micro base stations. In: ICC, pp. 1–6. Citeseer (2010)
21. Shafi, M., Molisch, A.F., Smith, P.J., Haustein, T., Zhu, P., Silva, P., Tufvesson, F., Benjebbour, A., Wunder, G.: 5g: A tutorial overview of standards, trials, challenges, deployment, and practice. IEEE J. Sel. Areas Commun. **35**(6), 1201–1221 (2017)
22. Siddique, U., Tabassum, H., Hossain, E., Kim, D.I.: Wireless backhauling of 5g small cells: challenges and solution approaches. IEEE Wirel. Commun. **22**(5), 22–31 (2015)
23. Tran, T.X., Hajisami, A., Pandey, P., Pompili, D.: Collaborative mobile edge computing in 5g networks: New paradigms, scenarios, and challenges. IEEE Commun. Mag. **55**(4), 54–61 (2017)
24. Xu, J., Wang, J., Zhu, Y., Yang, Y., Zheng, X., Wang, S., Liu, L., Horneman, K., Teng, Y.: Cooperative distributed optimization for the hyper-dense small cell deployment. IEEE Commun. Mag. **52**(5), 61–67 (2014)
25. Yunas, S.F., Valkama, M., Niemelä, J.: Spectral and energy efficiency of ultra-dense networks under different deployment strategies. IEEE Commun. Mag. **53**(1), 90–100 (2015)

ZoBe: Zone-Oriented Bandwidth Estimator for Efficient IoT Networks

Raghunath Maji, Souvick Das and Rituparna Chaki

Abstract IoT is made up of heterogeneous networks which transport a huge volume of data packets over the Internet. Improper utilization of bandwidth or insufficient bandwidth allocation leads to faults such as packet loss, setting up routing path between source and destination, reduction of speed in data communication, etc. One of the vital causes of insufficient bandwidth is nonuniform growth in the number of Internet users in a specific region. In this paper, we propose a framework for efficient distribution of bandwidth over a region based on depth of field analysis and population statistics analysis. We propose to use existing Google Earth Pro APIs over satellite images to estimate possible number of users in a particular area and plan to allocate bandwidth accordingly. The proposed framework is aimed to reduce packet loss and distortion effects due to scattering and refraction.

Keywords API · IoT · GIS

1 Introduction

Internet of Things (IoT) has opened up several exciting options to make our everyday living more advantageous and hence satisfying. Internet of things has caused a tremendous growth in Internet usage as most of these applications assume seamless connectivity to high-speed data network. The entire communication network is experiencing a larger-than-expected data requirement as all the devices, objects, and even humans have started to be data sources. In order to maintain the QoS metrics related to IoT, we are evolving new technologies for data communication. One of the issues in implementing these new technologies is improper bandwidth utilization, leading to inefficient data communication. Sudden growth of wireless users in a particular

R. Maji (✉) · R. Chaki
A. K. Choudhury School of Information Technology, University of Calcutta, Kolkata, India
e-mail: writetoraghunath@gmail.com

S. Das
Department of Computer Science and Engineering, University of Calcutta, Kolkata, India

© Springer Nature Singapore Pte Ltd. 2020   27
R. Chaki et al. (eds.), *Advanced Computing and Systems for Security*,
Advances in Intelligent Systems and Computing 995,
https://doi.org/10.1007/978-981-13-8962-7_3

area consumes higher bandwidth which can exceed the limit of the allocated bandwidth. In order to arrive at a proper forecast of data requirements and thus a proper planning for bandwidth distribution, an accurate zone-wise population is required.

Another problem of low signal strength in the ground floors of congested urban area is caused by improper height of BTS. Commercial telecom vendor-cum Internet service providers are interested in covering an area with minimum number of Base Transceiver Station (BTS) so that the spread spectrum is utilized properly.

In this paper, we propose a framework for population analysis in order to identify sudden growth of Internet users of a particular zone and measure the maximum required bandwidth for support.

Rest of the paper is organized as follows: Sect. 2 describes the state-of-the-art works, Sect. 3 describes the proposed methodology along with terminology, Sect. 4 presents a case study, and Sect. 5 concludes the paper.

2 Related Works

In this section, we present several existing works which have used Google Map, GIS (geographic information system), or satellite images for population analysis. In [1], authors proposed area calculation techniques for accurate estimation of rooftop area. This was done for efficient rainwater harvesting. In [2], authors proposed an automated digital image analysis framework that estimated the district population using satellite imagery. They have considered one satellite image as a training model and applied this model to the next image which is from a demographically similar area. The limitation of this paper is that, in case the next image is of a high-density area, then the population is underestimated and if low then overestimated. In [3], authors extracted building information from satellite image for proper estimation of the population. Though the previous census data has been used quite effectively, still, the authors failed to consider the building type that is not categorized. In [9], authors described a technique of measuring the land area using Google Earth API. The authors detect the longitude and latitude of the boundary region using geographic information system (GIS) and global positioning system (GPS). These are then used to calculate the inner land area without any human intervention. In [5], authors proposed a technique to identify the flood-prone district using satellite images data. The mechanism identifies a borderline between river and land. It also points out longitudes and latitudes of the borderline and identifies near districts that are most flood-prone areas. This mechanism uses the rational function model and rigorous sensor model for generating the ortho-images. In [6], authors use large-scale high ground resolution image and house count for estimating the population and also use different types of areas like residential, commercial, industrials, etc. but they cannot describe how to identify different types of areas. In [8], authors proposed a deep learning approach over previous census data and convolutional neural network (CNN) technique to predict the population of a specific area. In [4], authors aimed for accurate display positions of advertisements for student admission. The mechanism finds out the longitude and latitude of admitted student address using Google Earth API and calculates the distance from institute or college using Haversine formula. The distance calculation mechanism determines the appropriate position where the institute or college advertises for improving the admission rate. In [10], authors proposed a regional population estimation algorithm using satellite image. They used previous census data and building information. However, they did not consider the differences in building types (high rise, bungalows, etc.), and hence population density is estimated to be the same for all buildings.

3 Proposed Framework

In Fig. 1, the proposed framework is depicted. Different modules are described in the following section.

- **Keyword Parser**: Keyword parser is the module that parses a chunk of Google search keyword. It generates each keyword and feeds into another module called *keyword validator*.
- **Keyword Validator**: This module accepts those generated keywords from the *Keyword Parser* module and recognizes valid region searching keywords. It uses a valid list of keywords for regions that contain several keyword lists ordered by category. One example of the category can be *Educational Regions* which includes school, college, educational institutes, etc.
- **Building Identifier**: Based on the valid keyword, this module identifies different building complexes and other specified regions(slum) of a particular area. This entire module is powered by Google Earth APIs. We also extract several information about those building complex and regions with the help of decorated API calls.
- **Building Area Calculator**: We extract height, width, and coordinates of a building from the Google Earth API, and calculate the total area of the building using our algorithm mentioned in Algorithm 1.
- **Building-Type Classifier**: This module is responsible to classify the building type. It classifies buildings as educational sector, office complex, marketplaces, etc. This module helps to calculate population and the bandwidth requirements for a particular zone. The idea is that the bandwidth requirements for a office complex and market place will not be similar. This distinction is carried out based on the type of zone the module is classifying.
- **Population Calculator**: Based on the area of building and as per the category of the region, we calculate the population for that particular region.
- **Bandwidth Calculator**: Final module calculates the maximum bandwidth requirements for the entire area by summing up all regions in that area. It is worth mentioning that the maximum requirement of bandwidth for each node is 10gbps for 5G network [11].

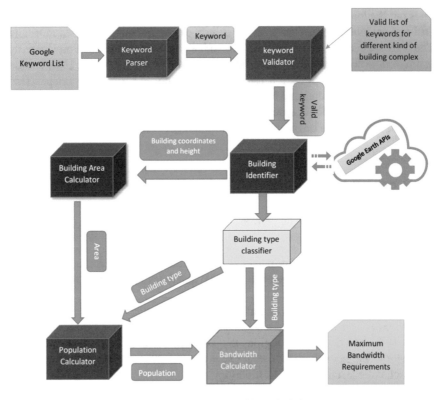

Fig. 1 The proposed framework for maximum bandwidth calculation

Figure 2 shows how data flow in the system and different sources of data. We briefly present how the framework actually works.

We have a bunch of search keywords from Google Earth to find different landmarks such as school, college, different institutes, organizations, industrial plants, hospital, restaurant, and so on. The interesting part is that some these search keywords are not mentioned in the straightway rather the name of the organization, institutes, colleges, and schools close to the regional languages, for example, an eye hospital named as "*Sankara Nethralaya*" and a school named as "Navodaya Vidyalaya". So we have maintained a list of those possible keywords. The module **Keyword Parser** parses necessary keywords where people can live or work. It will not accept keyword such as *water tank, road junction*, etc. The **Keyword Validator** module has two functionalities. One important aspect of the Google Earth API is that we can extract the type or category of a searched object. But we observed that categorization for several objects cannot be extracted using the Google Earth API. In that case, we have to use our predefined list of valid keywords. **Keyword Validator** does the extraction of categorization of the object and searches for a match in the valid keyword list. If the module is unable to extract categorization, then it searches for a match with the listed keywords with the words in its name. At this point, a valid identification and categorization of an object is completed. The next module called **Building Identifier** identifies and points the location of the building complex using Google Earth APIs. Moreover, this module extracts the height and width of the building from the APIs. The module **Building Area Calculator** calculates the area of the building based on our algorithm mentioned in Algorithm 1. It is worth mentioning that we are assuming the height of each floor of a building is about 11 ft. We create a polygon from the boundary coordinates of a building and in the next phase we use the **shoelace** [7] theorem to calculate the area from the polygon. $\mathbb{C}$ represents the set of coordinates of the boundary points of a building. So $\mathbb{C} = \{(x_1, y_1), (x_2, y_2), ..., (x_n, y_n)\}$. $\mathbb{H}$. Now we take a function $\mathscr{F}(\mathbb{C})$ for shoelace theorem which accepts $\mathbb{C}$ as its parameter. The function can be defined as follows:

$$\mathscr{F}(\mathbb{C}) = \frac{1}{2} \left| \sum_{i=1}^{n-1} x_i y_{(i+1)} + x_n y_1 - \sum_{i=1}^{n-1} x_{(i+1)} y_i + x_1 y_n \right|$$

Population Calculator module calculates approximate population for a particular area by calculating population for individual buildings in that area. The population calculation for each building is done based on statistical data like school, college admission rate, approximate employee strength for an industrial plan, and so on. It also estimates number of the Internet users. For example, in school the number of Internet users is much less though the population is considerably high. It is worth mentioning that **Population Calculator** module also estimates the time of sudden rise of population for an area and sudden fall of population of an area. For example, in an industrial hub area, there is a sudden fall of Internet users after 9 PM at night. Finally, the module **Bandwidth Estimation** calculates the maximum required bandwidth for that area based on the number of Internet users. Both population calculation and bandwidth estimation are carried out based on the building classification done by **Building-Type Classifier**. It classifies buildings as office complex, educational sector, marketplaces, etc.

Algorithm 2 Area Calculator

Input: $\mathbb{T}$ is the height of each floor of a building. $\mathbb{C}$ is the set of coordinates of a polygon that represents the border area of a particular building extracted from Google Earth API. The set $\mathbb{C}$ consists of latitudes and longitudes denoted as x_i and y_i, respectively. So $\mathbb{C} = \{(x_1, y_1), (x_2, y_2), ..., (x_n, y_n)\}$. $\mathbb{H}$ is the height of the building extracted from Google Earth API.

Output: $\mathbb{TA}$ is the total floor area generated for a particular building based on the coordinates.

1: **procedure** AREA CALCULATION
2:     Area of Floor($\mathbb{AF}$) $\leftarrow \mathscr{F}(\mathbb{C})$
3:     $\mathbb{NF} \leftarrow \mathbb{H}/\mathbb{T}$
4:     $\mathbb{TA} \leftarrow 0$
5:     **for** i=0 to $\mathbb{NF}$ **do**
6:         $\mathbb{TA} \leftarrow \mathbb{TA} + \mathbb{AF}$
7:     **end for**
8: **end procedure**

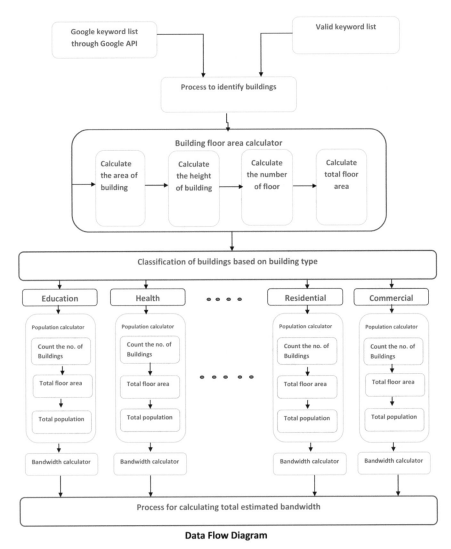

Data Flow Diagram

Fig. 2 Data flow for the proposed methodology

4   Case Study

In this section, we demonstrate our proposed logic and framework as a case study on University of Calcutta campus in Saltlake, India. We use Google Earth satellite image and various APIs from the Google Earth to evaluate our proposed logic. We have taken the Google Earth image of University of Calcutta, Saltlake campus building and mark all boundary points along with latitude and longitude to make a polygon.

Fig. 3 Area calculation using coordinate value from boundary points of University of Calcutta campus in Saltlake, India

Fig. 4 Calculating the height using Google Earth API

In the next phase, we calculate the area of the polygon using our proposed algorithm depicted in Sect. 3.

Figure 3 presents the selection of the boundary point and creation of a polygon of the satellite image of University of Calcutta, Saltlake campus. In this stage, we

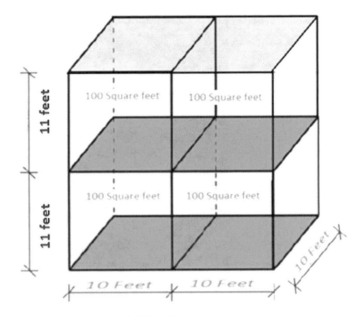

Fig. 5 Calculating the area of total building floor area

need to find out the number of floors of a building. In this context, we assume that the height of each floor of the building is about 11 ft. Now we extract the total height of the building from the Google Earth API.

At next, we calculate the number of floors of the building by simply dividing the total height by 11. We assume that the height of each floor of a building is 11. Generally, height of a floor is around 11 ft. Figure 4 depicts the extraction of height of the building from the Google Earth image.

In Fig. 5, we calculate the floor area for each floor using the previously mentioned Shoelace formula. The total floor area of the building is evaluated by simply summing up the floor area of every floor.

Based on region type, we estimate the population density. After that, we calculate the total population of the campus region based on the population density (Table 1).

5 Bandwidth Estimation

In Sect. 3, we proposed an algorithm to find the area calculation based on the height and width of a building. We also considered the number of floors within a particular building. In this section, we are more interested to approximate the total bandwidth consuming user for a specific region. We assume the population densities for different regions. For example, the density of bandwidth consumer for a residential complex is much less than the corporate sectors or other office premises. We also

Table 1 Actual implementation of total floor area calculation

Zone/area	Boundary Coordinates extracted from google API		Area of a floor extracted from google API	Height of building	Total floor area	Actual floor area	Error
	Longitude	Latitude					
Educational zone/ Calcutta university technology campus	88.41588736	22.56728348	29803 sq. ft.	55.65 ft.	149015 sq. ft.	30246 sq. ft.	443 sq. ft.
	88.41597587	22.56751135					
	88.4158659	22.56755097					
	.	.					
	.	.					
	.	.					
	88.41588736	22.56759556					
	88.41584176	22.56725871					
	88.41588736	22.56728348					

consider different aspects of calculation of bandwidth consumers. For example, we ignore the population densities in primary schools or upper primary schools as the growth of population in these areas is due to the growth of students. On the other hand, we consider the whole population as the bandwidth consumers. We have done some manual surveys to accumulate data for bandwidth consumers from different organizations like school, college, shopping mall, hospital, IT industry premises, and so on. Based on these data, we can calculate total bandwidth consumers for a particular region by considering each building complex and premises with their height and width measurements.

6 Conclusion and Future Work

In this paper, we calculate bandwidth requirements of an area from the calculation of total population of that area. We introduced a framework that measures population of an area by finding human residents, workplaces, etc., using Google Earth API. We also introduced an algorithm to calculate area of a building from the information retrieved from Google Earth API. Finally, we measure the population of the entire area by considering human inmates in individual building, complex, and marketplace. We do not use census data; the reason is that census data are fully human interpreted and error prone. Census data is some kind of static in nature. It only records household members. There is no data recorded for patients in hospitals. Census data is not well enough for dynamic analysis of an area such as railway station, industries, marketplaces, etc. We resolve this problem and enable minimum human intervention. We do not have to do any kind of survey to calculate area of a region; rather, calculation is done automatically. Thus, our framework measures the maximum required bandwidth for an area in very short time with a very low cost.

One limitation of our framework is that it cannot recognize abandoned buildings. Another drawback is that the framework estimates the inmates in a building which may not be the exact number. Our next work should be resolving these problems and build a tool for the entire framework.

We have evaluated our case study by manual execution of the different components required to support our proposed methodology. At this point, the implementation of the proposed methodology with a full functional tool support should be our next aim. On the other hand, extensive experiments should be carried out to understand the accuracy of the methodology.

References

1. Aher, M., Pradhan, S., Dandawate, Y.: Rainfall estimation over roof-top using land-cover classification of google earth images. In: 2014 International Conference on Electronic Systems, Signal Processing and Computing Technologies (ICESC), pp. 111–116. IEEE (2014)
2. Harvey, J.: Estimating census district populations from satellite imagery: some approaches and limitations. vol. 23, pp. 2071–2095. Taylor & Francis (2002)
3. Haverkamp, D.: Automatic building extraction from ikonos imagery. In: Proceedings of the ASPRS 2004, Annual Conference. Citeseer (2004)
4. Hegde, V., Aswathi, T., Sidharth, R.: Student residential distance calculation using haversine formulation and visualization through googlemap for admission analysis. In: 2016 IEEE International Conference on Computational Intelligence and Computing Research (ICCIC), pp. 1–5. IEEE (2016)
5. Jun, J.N., Seo, D.C., Lim, H.S.: Calculation of agricultural land flooding disaster area by typhoon from kompsat-1 eoc satellite image data. In: Proceedings 2004 IEEE International Geoscience and Remote Sensing Symposium, 2004. IGARSS'04, vol. 7, pp. 4678–4681. IEEE (2004)
6. Ogrosky, C.E.: Population estimates from satellite imagery **41**, 707–712 (1975)
7. Pure, R., Durrani, S.: Computing exact closed-form distance distributions in arbitrarily-shaped polygons with arbitrary reference point **17**, 1–27 (2015)
8. Robinson, C., Hohman, F., Dilkina, B.: A deep learning approach for population estimation from satellite imagery. In: Proceedings of the 1st ACM SIGSPATIAL Workshop on Geospatial Humanities, pp. 47–54. ACM (2017)
9. Setiawan, A., Sediyono, E.: Using google maps and spherical quadrilateral approach method for land area measurement. In: 2017 International Conference on Computer, Control, Informatics and its Applications (IC3INA), pp. 85–88. IEEE (2017)
10. Sutton, P., Roberts, D., Elvidge, C., Baugh, K.: Census from heaven: An estimate of the global human population using night-time satellite imagery. vol. 22, pp. 3061–3076. Taylor & Francis (2001)
11. Wang, C.X., Haider, F., Gao, X., You, X.H., Yang, Y., Yuan, D., Aggoune, H., Haas, H., Fletcher, S., Hepsaydir, E.: Cellular architecture and key technologies for 5g wireless communication networks. vol. 52, pp. 122–130. IEEE (2014)

Part II
Software Engineering and Formal
Specification for Secured Software Systems

Extracting Business Compliant Finite State Models from I* Models

Novarun Deb, Nabendu Chaki, Mandira Roy, Surochita Pal
and Ankita Bhaumick

Abstract Goal models are primarily used to represent and analyze requirements at an early stage of software development. However, goal models are sequence agnostic and fall short for analyzing temporal properties. Limited works are found in the existing literature that aims to bridge this gap. There are tools that transform a goal model to a finite state model (FSM). The existing works and their implementations can only check whether a given temporal property is satisfied by a goal model or not. However, it does not provide us with a compliant FSM which satisfies the compliance rules. This paper aims to generate a business compliant FSM for a given goal model specification that complies with the business rules (specified in some temporal logic). We have chosen to work with tGRL (textual modeling language for goal-oriented requirement language) as the goal model specification language for representing i* models. The framework extends the current i*ToNuSMV ver2.02 tool by allowing the user to give a CTL property as input along with a goal model. The proposed framework generates a compliant FSM that satisfies the CTL constraint.

N. Deb (✉) · N. Chaki · M. Roy · S. Pal · A. Bhaumick
University of Calcutta, Kolkata 700106, WB, India
e-mail: novarun.deb@unive.it

M. Roy
e-mail: roy.mandiracs@gmail.com

S. Pal
e-mail: pal.surochita@gmail.com

A. Bhaumick
e-mail: ankitabhaumik6@gmail.com

N. Deb
Università Ca' Foscari, Via Torino, 153, Venezia 30172, VE, India
e-mail: nabendu@ieee.org

© Springer Nature Singapore Pte Ltd. 2020
R. Chaki et al. (eds.), *Advanced Computing and Systems for Security*,
Advances in Intelligent Systems and Computing 995,
https://doi.org/10.1007/978-981-13-8962-7_4

1 Introduction

Goal-oriented requirements engineering (GORE) aims at providing the *rationales* for the requirements, which plays an important role in justifying the system to be developed [10, 12]. GORE formulates the initial requirements in such a way that it develops a deep understanding of the problem domain, interests, priorities, and abilities of various stakeholders [2]. Goal modeling frameworks have been developed for modeling these requirements. A goal is state of affairs that the system under consideration should achieve. i* [13] is one such goal-oriented requirements modeling notation that models requirements with the help of actors and their goals, tasks, and resources. Inter-actor dependencies are also captured in the i* framework. Creating a goal model to frame the requirements alone does not provide completeness to the whole idea of GORE. Goal models need to be analyzed to ensure their correctness and compliance to business rules. Although different types of goal model analysis techniques exist, yet goal models are inherently sequence agnostic which make them prone to errors arising out of noncompliance toward business properties and constraints. The i* goal modeling framework is no different and, in order to perform model checking on such models, we need to transform an i* model to a finite state model (FSM).

The existing i*toNuSMV ver2.02 tool [4] tries to bridge the gap by taking an input i* goal model (represented using tGRL [1]) and converts it into a finite state model (FSM), which can be mapped to the NuSMV input language. The output of the tool can be directly fed into the NuSMV model verifier and can be checked against temporal properties or compliance rules written using computational tree logic (CTL). The tool returns true if the property is satisfied (or fulfilled); otherwise, it provides a counterexample (where the property is not satisfied). In reality, there may exist more than one *execution paths* within the finite state where a particular temporal property may fail to satisfy, which the model checker does not provide us.

The main objective of the framework presented here is to prune all those execution paths from the FSM and generate a modified FSM that is compliant to the CTL property. The proposed tool goes one step further and prunes all those state transitions from the FSM (generated by the existing i*ToNuSMV version 2.02) that result in constraint violations. The extended version of the tool identifies the particular combinations of leaf-level goal, task and resource fulfillment that results in the contradiction of the given CTL constraint. This allows the proposed version of the tool to derive a pruned finite state model that satisfies the compliance constraint in the requirement phase itself. It aims at providing a deep and meaningful understanding of each of the requirements modeled using i* and all the situations where the requirements may or may not be fulfilled. This may be particularly helpful for system designers in developing a nonambiguous system.

A knowledge of the existing i*ToNuSMV ver2.02 tool is a pre-requirement for understanding the extension provided to the tool in this paper. The URL for accessing the existing version 2.02 of the tool is as follows: http://cucse.org/faculty/tools/. We will assume that the reader has a sound knowledge of the working of the current tool version (ver2.02) and skip the discussion of some of the basic concepts which have been inherent in all previous versions of the tool. For a brief understanding of the working of the current version of the tool, readers can refer to the user manual given in this link: http://cucse.org/faculty/wp-content/uploads/2017/03/User-Manual_ver2.02.pdf. The rest of the paper is organized as follows: Sect. 2 presents a brief review of the current state of the art for ensuring business compliance in goal modeling. In Sect. 3, we present the basic methodology and assumptions underlying this tool extension. This is followed by Sect. 4 where we elaborate with the help of use cases, the different types of CTL properties that we have addressed in this tool extension. The section also presents an algorithm for generating the tool. Finally, we conclude the paper with Sect. 5.

2 Review

Business processes undergo evolutionary change of lifecycle from an unsatisfactory state to desired state. This volatility of business process models presents the need of methods to control and trace the evolution process [9]. There are works that represented business process model in a goal model form and specifications are applied to the goal model to check the validity of the process model. The addition, removal, or modification of goals ultimately result in modification of temporal ordering of events. Business processes and their management have always introduced challenges for organizations, as the processes are often cross-functional. In [11], the author has proposed a business process management methodology and framework based on the User Requirements Notation as the modeling language and jUCMNav as the supporting tool for modeling and monitoring processes. Business processes are subject to compliance rules and policies that stem from domain-specific requirements such as standardization or legal regulations. There are instances when the constraint that has to be applied on process data rather than on the activities and events, so checking such constraint should deal with data conditions as well as ordering of events. Knuplesch et al. [8] have proposed a data-aware compliant checking methodology as a preprocessing method to enable correct verification. A methodology that directly transforms a business process model into finite state machine and performs compliant checking using linear temporal logic (LTL) has been proposed in [7]. It is mainly focused on verification of control flow aspects rather than verification of the state of data objects. Reference [10] is focused on the fact that deployment efficiency of business process models can be improved by checking the compliance of business process models by model checking technology. The existing literature [10] in this domain highlights the importance of model checking against temporal properties and has come with various ways to do so.

Often business process models are represented using goal models for validating the requirements. Unlike dataflow models, goal models like i* are sequence agnostic which make them prone to errors arising out of noncompliance toward temporal properties. Works have been done [3] to bridge this gap by transforming goal models into finite state models. These finite state models can be fed into standard model verifiers like NuSMV and can be checked against temporal properties. The current i*toNuSMV ver2.02 can only check a given goal model against temporal properties [4]. However, if the property is not satisfied it does not identify the state transitions due to which the property fails to get satisfied. Several goal model analysis techniques have been proposed in recent years, some techniques propagate satisfaction values through links to and from goals in the model, others apply metrics over the structure of the model, apply planning techniques using tasks and goals in the model, run simulations over model tasks, and yet others perform checks over model contents [6]. A survey of these different approaches shows that model checking using temporal property provides satisfactory result, but an iterative process of manually defining the bounds of the model checker is often required.

In [5], the authors have introduced a goal-oriented requirements engineering method for compliance with multiple regulations. Legal statements are modeled using Legal-GRL and their solutions for a goal model have been linked with organizational-GRL. The business processes and their operations should satisfy a set of policies or constraints characterized by compliance rules. The change in compliance rules can occur in line with the business goals, and also with legal regulations. Reference [12] has highlighted the importance of legal compliance checking in business process model. Any noncompliance may lead to financial and reputation loss. A review of various goal-oriented frameworks for business compliance checking indicates that more research in this domain is required to face different challenges that come up with changing compliance rules. The existing framework of model checking only provides a counterexample when compliance rules are not satisfied but it does not come up with the exact model that is compliant with the given temporal property.

3 Methodology

In this section, we briefly describe some of the basic concepts of the existing version of the tool with the help of which we will elaborate our proposed framework in the following section.

3.1 The I*ToNuSMV FSM

The finite state model generated is a state transition model in which each identifier (associated with every goal model element) goes through three possible states: *0(Not Created—NC), 1(Created Not Fulfilled—CNF), 2(Fulfilled—FU)*. When a new model element (goal, task, resource) is encountered, the corresponding system variable $V\#$(# be the integer value assigned) is initialized with 0 representing *Not Created* state. A model element is said to be fulfilled if it has no child element or its child element(s) are individually fulfilled. A model element can be dependent on its child node(s) in two ways: task decomposition and means-end decomposition. A *task decomposition* is an AND decomposition and requires that all the child elements must be fulfilled in order to declare the parent node fulfilled. A *means-end decomposition* is an OR decomposition and provide alternate strategies to fulfill the parent node element.

3.2 Example

Considering a goal model where an actor A has a goal G_1 and G_1 has three subgoals (child elements) G_2, G_3, and G_4 which are connected to G_1 by means-end decomposition. The goal tree for the actor A and the finite state model (generated using Semantic-Implosion algorithm [3]) for the goal model is shown in (Fig. 1a).

Let us consider a CTL constraint—EF (G_1=FU). Based on the semantics of the EF CTL operator, i*ToNuSMV ver2.02 tries to find an execution trace within the generated FSM that satisfies this constraint. The model checker traces the state

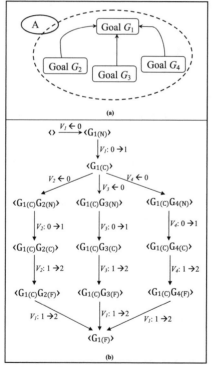

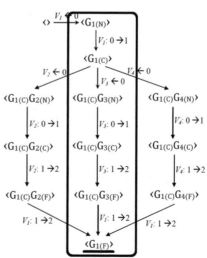

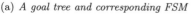

(a) *A goal tree and corresponding FSM*   (b) *The state and path in FSM where CTL constraint holds true*

Fig. 1 Constraint checking in OR decompositions [3]

transitions of the finite state model and finds one of the paths (as shown in Fig. 1a) where the constraint holds true.

In this particular example, even the CTL constraint $\text{AF}(G_1{=}\text{FU})$ would hold true as all paths eventually lead to a state where goal G_1 is fulfilled (or, $V_1 = 2$). However, if we try to check the CTL constraint $\text{EG}(G_1!{=}\text{FU})$, then the NuSMV model verifier will return false because in every path the last state results in fulfilling G_1. The research question being answered in this paper is that—"*How do we modify the generated FSM such that the property* $\text{EG}(G_1!{=}\text{FU})$ *is satisfied?*"

3.3 Assumptions

We have thoroughly studied the different types of CTL constraints and have worked with a finite subset of such constraints in our tool. Our primary goal is to provide a pruned finite state model (generated by i*ToNuSMV ver2.02) that satisfies the compliance rules. This work has the following four assumptions:

A-1: Since FSMs are derived for fulfillment of goals, we work with only AG and EG temporal operators for the violation of goal fulfillment. Example:$\text{AG}(\text{V109}!{=}\text{FU})$.

A-2: Two CTL predicates can be connected through Boolean connectives like AND and OR. Our tool allows the user to define only two predicates at a time and connect them by theAND or OR operator. Example: $\text{AG}(\text{V109}!{=}\text{FU AND}$ $\text{V102}!{=}\text{FU})$.

A-3: Another type of CTL constraint that we have worked with is implication ($\rightarrow$). Any two constraints can have implication between them. We have restricted implication to only single level of nesting. Example: $\text{AG}(\text{V101}{=}\text{CNF}\rightarrow$ $\text{AF}(\text{V102}{=}\text{FU AND V103}!{=}\text{FU}))$.

A-4: We have assumed the goal tree level for an actor to be three to reduce the problem complexity.

We will provide a detailed explanation about each of the CTL constraint that we have worked with and how we derive the different finite state models in the following section.

4 Case Study

4.1 *EG(V#!=FU) for and Decomposition*

Consider an i* model (Fig. 2a) consisting of a goal G_1 (V101) whose fulfillment depends upon the child goals G_2 (V102) and G_3 (V103). The decomposition type for goal G_1 is *task decomposition*, which means G_1 will be fulfilled *iff* both its child goals G_2 and G_3 are achieved.

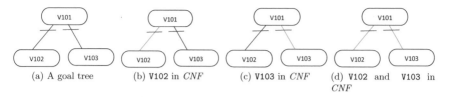

(a) A goal tree   (b) V102 in *CNF*   (c) V103 in *CNF*   (d) V102 and V103 in *CNF*

Fig. 2 Different paths through which constraint is satisfied

Given a CTL constraint EG(V101!=FU), which has to be applied on the goal model (in Fig. 2a). The CTL formula V101!=FU refers to the state where the goal V101 is not fulfilled. EG refers to a path where the constraint is satisfied globally. The goal V101 fails to get fulfilled if any of its child goal V102 and V103 are not achieved or both of them are not achieved (since it is an AND decomposition). There will be three possible ways to generate a compliant finite state model:

1. The goal V102 is not achieved, i.e., it remains in *CNF* state, which will violate the AND decomposition criteria and so V101 will not be fulfilled (Fig. 2b).
2. The goal V103 is not achieved, i.e., it remains in *CNF* state, which will violate the AND decomposition criteria and so V101 will not be fulfilled (Fig. 2c).
3. The goals V102 and V103 both are not achieved, i.e., both remains in *CNF* state, which will also violate the AND decomposition criteria and so V101 will not be fulfilled (Fig. 2d).

Each of these situations will be represented in a separate finite state model. The finite state models will be created in such a way that the child goals V102 and V103 do not make transition to fulfilled state, i.e., 1->2 (following each of the case mentioned).

4.2   *EG (V#!=FU) for OR Decomposition*

Consider an i* model (Fig. 3a) consisting of a goal G_1(V109) whose fulfillment depends upon the child goals G_2 (V110) and G_3 (V111). The decomposition type for goal G_1 is *means-end decomposition*, which means G_1 has alternative strategies for getting fulfilled, i.e., either when child goal G_2 is achieved or when G_3 is achieved or both.

Given a CTL constraint EG(V109!=FU), which has to be applied on the goal model (in Fig. 3a). The CTL formula V109!=FU refers to the state where the goal V109 is not fulfilled. EG refers to a path where the constraint is satisfied globally. The goal V109 fails to get fulfilled when none of its alternative child goals (V110 and V111) is achieved. So there will be only one possible way to generate finite state model:

Fig. 3 Path through which
constraint is satisfied

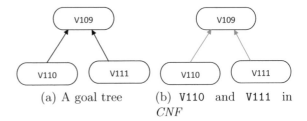

(a) A goal tree   (b) **V110** and **V111** in
                                                      CNF

1. The child goals V110 and V111 both are not achieved, i.e., both remains in *CNF* state, which will violate the OR decomposition criteria and so V109 will not be fulfilled (Fig. 3b).

   The finite state models will be created in such a way that the child goals V110 and V111 do not make transition to fulfilled state, i.e., 1->2.

4.3 EG(V#!=FU and V#!=FU)

We consider an i* model consisting two goal models of two different actors (Figs. 2a and 3a). Given a CTL constraint EG(V101!=FU AND V109!=FU), which has to be applied on the abovementioned multiple-actor i* model. The CTL formula V101!=FU and V109!=FU refers to the state transitions where the goals V101 and V109 are not fulfilled. Such a scenario will take place *iff* for V101 either of the child goals or both are not achieved and for V109 it will occur only if none of the child goals are fulfilled. There are three possible finite state models (Fig. 2) that can be generated to satisfy EG(V101!=FU) and there is only one finite state model (Fig. 3b) that can be generated to satisfy EG(V109!=FU).

The two CTL predicates are joined by Boolean connective AND, so we need to generate finite state models where both the formulae satisfy simultaneously. Let us suppose that the first predicate can be satisfied in *M* ways and the second predicate can be satisfied in *N* ways. Since both must occur simultaneously, there are *M* × *N* ways to satisfy both (according to fundamental principle of counting theory). Thus, in this case, there will three different finite state models satisfying the constraint. Each of the possible solutions for first predicate is combined with the possible solution of the second predicate in the CTL constraint.

4.4 EG(V#!=FU or V#!=FU)

We again consider the single i* model comprising of the two goal models (shown in Figs. 2a and 3a). Given a CTL constraint EG(V101!=FU OR V109!=FU), which has to be applied on the multiple-actor i* model. The two CTL predicates are joined by Boolean connective OR, so we need to generate finite state models where either the first predicate is satisfied, or the second predicate is satisfied, or both. The number of finite state models generated would, therefore, be $M + N + (M \times N)$, where $M + N$ implies models where either of the predicates are satisfied and $M \times N$ implies that both predicates are satisfied. The OR and AND connective can also be applied for the identifiers of the same goal tree.

4.5 AG(V#!=FU→ V#!=FU)

Consider a goal model (Fig. 4a) where root goal G_1(V1) is task decomposed into two child goals G_2(V2) and G_3(V3), i.e., G_1 will be fulfilled *iff* both its child goals G_2 and G_3 are achieved. The corresponding finite state model, as extracted by i*ToNuSMV ver2.02, is also shown in Fig. 4a. The lattice structure represents all possible valid permutations for satisfying goals G_2 and G_3. Let us consider a CTL property AG(V2!=FU→V3!=FU) which has to be complied by the goal model considered. The property encodes a temporal ordering between the goals G_2 and G_3. In the given goal model, child goal G_2 should be fulfilled before child goal G_3. The CTL constraint demands that goal G_2 must be fulfilled before G_3. Since G_3 cannot be fulfilled any time before G_2, our proposed framework will prune all those state transitions from the FSM which result in a contradiction of the CTL constraint. The pruned FSM is shown in Fig. 4b.

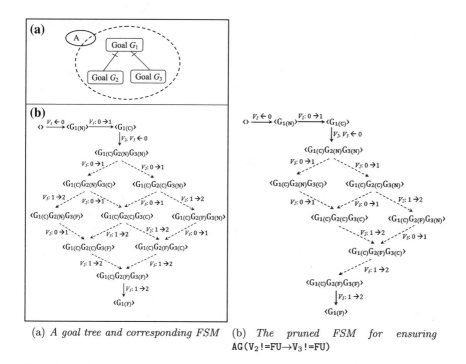

(a) *A goal tree and corresponding FSM*   (b) *The pruned FSM for ensuring* AG(V$_2$!=FU→V$_3$!=FU)

Fig. 4 Constraint checking for AND decomposition [3]

4.6 *EG(V#!=FU) for **AND–OR** Decomposition*

We consider an i* model (Fig. 5a) with parent goal G_1 (V109) which consists of two child goals G_2 (V110) and G_3 (V111). The decomposition type for parent goal is AND which means both G_2 and G_3 need to be fulfilled for goal G_1 to be achieved. The goal G_2 is means-end decomposed to task T_1 (V112) and T_2 (V113), i.e., G_2 has two alternative strategies for getting fulfilled. The goal G_3 is also means-end decomposed to task T_3 (V114) and resource R_1 (V115).

Given a CTL constraint EG(V109!=FU), we need to find all the different finite state models where parent goal V109 is not fulfilled. The parent goal V109 will not be fulfilled if either of its child goal or both are not achieved. The child goal V110 will not be achieved if none of the task (V112 or V113) is performed. Similarly, child goal V111 will not be achieved if neither task V114 is performed nor resource V115 is obtained. So, for each of the child node V110 and V111 we have only one possible finite state model to satisfy the constraint. The parent goal V109 will not be fulfilled if state transitions of left child remains in *CNF* state or state transitions of right child remains in *CNF* state or both remains in *CNF* at the same time. So we will have three different finite state models with different state transitions that satisfy the constraint (Fig. 5). The possible set of solutions are as follows:

1. Leaf-level tasks V112 and V113 are not performed, so child goal V110 is not fulfilled thus V109 also remains in not fulfilled state (Fig. 5b).
2. Leaf-level task V114 is not performed and resource V115 is not obtained, so child goal V111 is not fulfilled thus V109 also remains in not fulfilled state (Fig. 5c).
3. Leaf-level tasks V112, V113, and V114 are not performed and resource V115 not obtained, so child goals V110 and V111 are not fulfilled thus V109 also remains in not fulfilled state (Fig. 5d).

4.7 *EG(V#!=FU) for **OR–AND** Decomposition*

We try to illustrate the working CTL constraint using another complex i* model (Fig. 6(a)) with OR–AND decomposition. Considering an i* model with a root goal G_1 (V109) which is decomposed into two child goals G_2 (V110) and G_3 (V111).

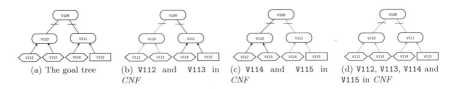

(a) The goal tree   (b) V112 and V113 in   (c) V114 and V115 in   (d) V112, V113, V114 and
                           CNF   *CNF*   V115 in *CNF*

Fig. 5 Different paths through which constraint is satisfied

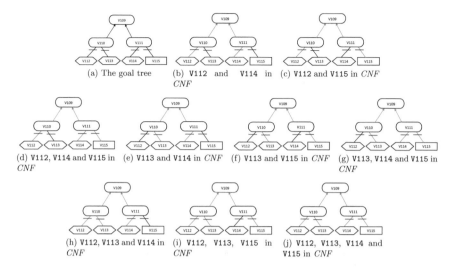

Fig. 6 Different paths through which constraint is satisfied

The decomposition type is means-end, i.e., G_1 has two alternative strategies for getting fulfilled. The child goal G_2 is further decomposed into two leaf-level tasks T_1 ((V112) and T_2 (V113). The decomposition type is task decomposition, i.e., G_2 can be fulfilled when both the tasks T_1 and T_2 are performed. The child goal G_3 is decomposed into task T_3 (V114) and resource R_1 ((V115)), which is also a task decomposition.

Given a CTL constraint EG(V109!=FU), we are required to find all different paths in the finite state model where the root goal V109 does not get fulfilled globally. A situation like this will arise when none of the child goals is fulfilled (through any of the path). The child goal V110 will not get fulfilled if either of the tasks (V112 or V113) is not performed or both. The child goal V111 will not get fulfilled if either of the tasks V114 is not performed or resource V115 is not obtained or both. Therefore, we can generate three possible finite state models for each of the child goals (V110 and V111) where the constraint will be satisfied.

The parent goal V109 will not be fulfilled when both its child goals are not fulfilled simultaneously. The final set of finite state models is generated by combining the solutions of the child nodes. Again by using counting theory there will be 3×3 different finite state models each depicting a different state transition path that satisfy the constraint (Fig. 6). The possible set of solutions are as follows:

1. Leaf-level tasks V112 and V114 are not performed (Fig. 6b).
2. Leaf-level task V112 is not performed and resource V115 is not obtained (Fig. 6c).
3. Leaf-level tasks V112 and V114 are not performed and resource V115 not obtained (Fig. 6d).
4. Leaf-level tasks V113 and V114 are not performed (Fig. 6e).

Algorithm 1 Finite State Model generation

Input: A goal model with a temporal property.
Output: A finite state model satisfying the temporal property.
Data Structure: A linked list storing the goal tree for each actor in the goal model.

1: Each model element (goal, task, resource) is assigned with a variable name. The variable mapping for each element is displayed in the tool.
2: A goal tree is generated from the given i* model.
3: The type of CTL constraint is identified
4: **if** Constraint is not implication type **then**
5:     **if** The constraint has a single variable **then**
6:         Locate the variable in the goal tree and check whether it is a leaf or non-leaf node.
7:         **if** The variable is a leaf node in the tree **then**
8:             Store the variable in file and go to Step 35
9:         **else**
10:             Store its decomposition type and locate the child nodes.
11:             **if** The child nodes are not leaf **then**
12:                 **for** Each of the child node **do**
13:                     Get the leaf-level child and combine them according to the decomposition type so as to satisfy the CTL constraint
14:                 **end for**
15:                 The leaf level child nodes combination are further combined according to the decomposition type of the variable in the constraint and stored in a file.
16:             **else**
17:                 Combine them according to the decomposition type of the variable located in CTL specification and store it in a file.
18:             **end if**
19:         **end if**
20:     **else if** The constraint has a Boolean connective **then**
21:         **for** Each variable in the CTL constraint **do**
22:             Do step 7 to 20
23:         **end for**
24:         Check the Boolean connective used in the CTL constraint and combine the solution for each of the variable accordingly and store it in the file.
25:     **end if**
26: **else if** There is an implication operator **then**
27:     Check whether implication will hold true for the given goal model.
28:     **if** Implication holds true **then**
29:         Store the variable in a file whose state transitions need to be modified to make the implication true.
30:     **else if** Implication is not true **then**
31:         Generate appropriate message
32:     **end if**
33: **end if**
34: **for** Each combination of node stored in the file **do**
35:     Modify the state transitions of the nodes to satisfy the temporal property
36:     Generate separate finite state model for each solution path in the goal tree.
37: **end for**

5. Leaf-level task `V113` is not performed and resource `V115` is not obtained (Fig. 6f).
6. Leaf-level tasks `V113` and `V114` are not performed and resource `V115` not obtained (Fig. 6g).
7. Leaf-level tasks `V112`, `V113`, and `V114` are not performed (Fig. 6h).
8. Leaf-level tasks `V112` and `V113` are not performed and resource `V115` is not obtained (Fig. 6i).
9. Leaf-level tasks `V112`, `V113` and `V114` are not performed and resource `V115` not obtained (Fig. 6j).

Algorithm 1 provides a set of sequential steps required for generating the pruned finite state model satisfying a given CTL constraint.

5 Conclusion and Future Work

In this work, we present an initiative to generate finite state models from goal models that already satisfy some given CTL constraint. Finite state models can be more readily transformed into code. Thus, this research takes an important step toward the development of business compliant applications directly from goal models. Most business compliance rules have some sort of temporal ordering over events and can be represented with temporal logics efficiently. However, the proposed solution has several assumptions which need to be relaxed for making the framework more complete. We are also in the process of developing a tool interface called i*ToNuSMV ver3.01—the next version of the i*ToNuSMV tool—that will allow enterprise architects to use our proposed framework.

Acknowledgements This work is a part of the Ph.D. work of Novarun Deb, who was a Research Fellow in the University of Calcutta under the Tata Consultancy Services (TCS) Research Scholar Program (RSP). We acknowledge the contribution of TCS Innovation Labs in funding this research. Part of this work was done by Novarun Deb at the Decision Systems Lab, University of Wollongong. We acknowledge the Technical Education Quality Improvement Programme (TEQIP), University of Calcutta, for organizing and sponsoring his visit to the university in Wollongong, Australia. This work has also been partially supported by the Project IN17MO07 "Formal Specification for Secured Software System", under the Indo-Italian Executive Programme of Scientific and Technological Cooperation.

References

1. Abdelzad, V., Amyot, D., Alwidian, S.A., Lethbridge, T.: A Textual Syntax with Tool Support for the Goal-oriented Requirement Language. University of Ottawa, Ottawa, Canada, EECS (2015)
2. Deb, N., Chaki, N., Ghose, A.K.: Using i* model towards ontology integration and completeness checking in enterprise systems requirement hierarchy. In: IEEE International Model-Driven Requirements Engineering Workshop (MoDRE) (2015)

3. Deb, N., Chaki, N., Ghose., A.K.: Extracting finite state models from i* models. J. Syst. Softw. **121**, 265–280 (2016). https://doi.org/10.1016/j.jss.2016.03.038
4. Deb, N., Chaki, N., Ghose, A.K.: i*tonusmv: a prototype for enabling model checking of i* models. In: IEEE 24th International Requirements Engineering Conference (RE) (2016)
5. Ghanavati, S., Rifaut, A., Dubois, E., Amyot, D.: Goal-oriented compliance with multiple regulations. In: IEEE 22nd International Requirements Engineering Conference (RE) (2014). https://doi.org/10.1109/RE.2014.6912249
6. Horkoff, J., Yu, E.K.H.: Analyzing goal models: different approaches and how to choose among them. SAC (2011)
7. Kherbouche, O.M., Ahmad, A., Basson, H.: Formal approach for compliance rules checking in business process models. In: IEEE 9th International Conference on Emerging Technologies (ICET) (2013). https://doi.org/10.1109/ICET.2013.6743500
8. Knuplesch, D., Ly, L.T., Rinderle-Ma, S., Pfeifer, H., Dadam, P.: On enabling data-aware compliance checking of business process models. In: 29th International Conference on Conceptual Modeling, Springer, vol. 6412, pp. 332–346 (2010). http://dbis.eprints.uni-ulm.de/665/
9. Koliadis, G., Ghose, A.K.: Relating business process models to goal-oriented requirements models in kaos. PKAW (2006)
10. Negishi, Y., Hayashi, S., Saeki, M.: Establishing regulatory compliance in goal-oriented requirements analysis. In: IEEE 19th Conference on Business Informatics (CBI) (2017). https://doi.org/10.1109/CBI.2017.49
11. Pourshahid, A., Amyot, D., Peyton, L., Ghanavati, S., Chen, P., Weiss, M., Forster, A.J.: Business process management with the user requirements notation. Electron. Commer. Res. **9**, 269–316 (2009)
12. Sepideh, G., Daniel, A., Liam, P.: A systematic review of goal-oriented requirements management frameworks for business process compliance. In: IEEE Fourth International Workshop on Requirements Engineering and Law (2011). https://doi.org/10.1109/RELAW.2011.6050270
13. Yu, E.: Modelling strategic relationships for process reengineering. University of Toronto, Toronto, Canada Ph.D. thesis (1995)

Behavioral Analysis of Service Composition Patterns in ECBS Using Petri-Net-Based Approach

Gitosree Khan, Anirban Sarkar and Sabnam Sengupta

Abstract The service composition and scheduling activities are facing performance and complexity issues because of (a) large number of heterogeneous clouds, (b) integrating various service components into composite service. In order to facilitate such issues, the advent of automatic dynamic web service composition and scheduling methodology is required, such that the current trends of problem like service reusability, flexibility, statelessness, efficiency, etc. can be addressed. This work focus on the web service composition process in multi-cloud architecture, where various types of service composition patterns are discussed. The service composition patterns are classified according to the degree of heterogeneity of the services. It also helps to design the dynamic facets of composition patterns using Enterprise Service Composition Petri net (ESCP) model. Further, using the concepts of ESCP model and the reachability graph, several key properties like safeness, boundedness, fairness, etc. are analyzed formally.

Keywords Enterprise cloud bus · Service composition · Behavioral analysis · Colored Petri net · Deadlock · Reachability

1 Introduction

The requirement for cloud computing technology in the areas of enterprise software applications has evolved continuously on a large scale that identifies functional requirements from end users keeping functional cost and service access. Many software enterprises are heading toward the development of cloud services and therefore

G. Khan (✉) · S. Sengupta
B.P. Poddar Institute of Management and Technology, Kolkata, India
e-mail: khan.gitosree@gmail.com

S. Sengupta
e-mail: sabnam_sg@yahoo.com

A. Sarkar
National Institute of Technology, Durgapur, India
e-mail: sarkar.anirban@gmail.com

© Springer Nature Singapore Pte Ltd. 2020
R. Chaki et al. (eds.), *Advanced Computing and Systems for Security*,
Advances in Intelligent Systems and Computing 995,
https://doi.org/10.1007/978-981-13-8962-7_5

it is difficult to manage services in a very large scalable and virtualized manner. Therefore, lots of problems arise in the service discovery, composition, and scheduling process that comprises complexity, performance, and scalability issues. These issues are addressed by establishing the concept of hierarchical layer called Enterprise Cloud Bus System (ECBS) [1]. The proposed architecture is composed of the components and their interconnections, and delivers the service to the end users. This methodology is implemented by integrating the concept of multi-agent technology. Some of our previous work is based on service registration, discovery, and scheduling of services [2, 3] in multi-cloud architecture. Furthermore, a high-level design of enterprise cloud bus architecture is established using Petri net tool [4].

However, there is no such proper mechanism to support all these theoretical concepts for verifying the service composition patterns in multi-cloud architecture. Therefore, this paper focuses on modeling and verification of service composition techniques using Petri-net-based approach, which compose cloud service dynamically based on degree of heterogeneity concepts. The novelty of the proposed service composition approach coordinates cloud participants, reconfigure scheduled services, and focuses on aggregating a composite services in inter-cloud environments using several service composition patterns as discussed in the paper. Besides, several behavioral properties of service composition patterns have been analyzed using the proposed Petri net model and are simulated using a case study. The motivation of the proposed work is to establish an agile and flexible collaboration among services, composing, and executing service workflows. The proposed framework is capable of composing the heterogeneous service and facilitates the functional aspects of service composition process in enterprise cloud-based applications in terms of flexibility, scalability, integrity, and dynamicity of the cloud bus.

2 Review of Related Work

The recent trends in growing enterprise toward service-oriented computing [5, 6] became one of the major reasons for accessing the computing resource in enterprise applications platforms in terms of scalability and flexibility. Few researchers [7] discuss the architectural design of cloud computing and its applications in business enterprise. The emergence of automatic web service composition process is required, such that the current trends of problem like service reusability, flexibility, statelessness, efficiency, etc. can be addressed due to increase in overhead of clouds and its services. Traditional web service composition [8] and scheduling [9] methodology is hard to support dynamic, service composition workflow because it does not have rapid monitoring strategies for error handling and dynamic reconfiguration. Thus, in paper [10], the author focuses on agent-based service composition approaches that facilitate the challenges of service composition pattern dynamically. Moreover, in [11], the author proposes a self-organizing multi-agent-based dynamic approach for service composition patterns. Furthermore, to make the service composition process more formal, many of the research work [12] focuses on the modeling and verification of service composition and scheduling [13] patterns using Petri-net-based approach. Paper [14] discusses some theoretical and experimental results based on heuristic search algorithm using timed Petri nets. Further, paper [15] presents a verification technique on web services composition approach using high-level Petri net concepts [4]. A high-level Petri net is defined as a directed bipartite graph which consists of two nodes, namely, places and transitions. The arc that connects the nodes represents state of a transition of the given node. In this paper, the research work [16] focus on MAS based modeling of a distributed system using Colored Petri-net tool.

However, several research works are discussed on formal modeling of service composition patterns, but still there are few limitations on modeling and verification of service composition patterns in inter-cloud environment. This paper focuses on modeling, analysis, and verification of dynamic service composition approach in ECBS using Petri-net-based model.

3 Service Composition Patterns

In this section, dynamic service composition patterns are described based on the degree of heterogeneity concepts. The degree of heterogeneity is measured in terms of service capability of the process present within the system. The process is defined as perceived sequence invocations of cloud services where the perceived sequence may be either structured (repeatable) or/and nonstructured (non-repeatable) patterns. To define various categories of cloud and its reachable combinations, here we have considered Cloud Service Index (CI), a metric that measures the total number of clouds and its associated service coming from different location contexts. Formally, the Cloud Service Index (CI) can be defined as

$$C_I = C_k * S_j$$

where $(C_k * S_j)$ indicates total number "kth" unit of cloud (C) combine with "jth" number of service (S) depending upon the availability of cloud services. In this section, we have discussed various metrics for defining the measurement of degree of heterogeneity in terms of Service Capability (SC_n). The Service Capability (SC_n) defines the difference between the Upper Capability (UCL) and Lower Capability (LCL) limit of the service divided by the standard deviations of predictive service functionality. Thus, the difference between the service capabilities is the difference between the numbers of process invoked by a service during its run time.

Therefore, Service Capability (SC_n) formally can be defined as

$$SCn = (UCL - LCL)/\sigma \quad (1)$$

where UCL is the upper capability limit, LCL is lower capability limit, and σ is the standard deviation of predictive service functionality.

In order to calculate the value of *UCL* and *LCL*, we have used the *P* chart concept, where n is the size of each cloud service S. Therefore,

$$UCL = P' + i\sqrt{\{P'(1 - P')/C_k * S_j\}} \qquad (2)$$

$$LCL = P' - i\sqrt{\{P'(1 - P')/C_k * S_j\}} \qquad (3)$$

where $i = 3$, which is a plier chosen to control the out of control line of the limits.

Thus, the Degree of Heterogeneity (H^2) formally can be defined as

$$H^2 = \sum_{k=1}^{m} C_k * \sum_{j=1}^{n} S_j * \left\{1/2 * \left(\sum (SC_{n2} - SC_{n1})/SC_n\right) * 100\%\right\} \qquad (4)$$

Figure 1 shows the graphical representation of service capability versus number of service participate in the service composition patterns.

In this work, the Lower Capability Limit (*LCL*) value is considered as 30 and Upper Capability Limit (*UCL*) value is considered as 60. It has been observed from the graph that the service having service capabilities less than *LCL* value follows choreography pattern because of its lower degree of heterogeneity and less coherence among the participants' service. In similar way, the services which comprise more than *UCL* values follow orchestration pattern because of high degree of interactivity among the services and the services having service capability between *LCL* and *UCL* values follow hybrid pattern of service composition because of moderate nature of degree of heterogeneity among participants service.

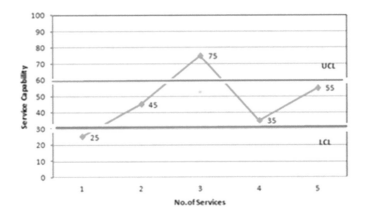

Fig. 1 Service capability versus number of services

3.1 Graphical-Based Model of Service Composition Patterns

In this section, different types of service composition patterns (service choreography, service orchestration, and service hybrid) are discussed on the basis of degree of heterogeneity concepts. The composition pattern of heterogeneous service is classified based on the following range of considerations mentioned in [13]:

Considering the range of heterogeneity, we can state that

(a) *If Degree of Heterogeneity lies between 0% and 30%, choose Choreography Pattern.*
(b) *If Degree of Heterogeneity lies between 30% and 60%; choose Hybrid Pattern.*
(c) *If Degree of Heterogeneity lies between 60% and 100%, choose Orchestration Pattern.*

The service choreography pattern has been chosen considering lower heterogeneity and lesser coherence of services (homogeneous service/homogeneous cloud) since the choreography pattern is considered to be decentralized, less flexible, less reliable, and modifiable. The service orchestration pattern is chosen considering upper heterogeneity or higher coherence among the service, because of its centralized pattern, more flexibility, and reliability and can change the process workflows service composition easier. While, in hybrid-based pattern, the services interact among each other moderately and have a high chance of interactivity among the services. The utility of *SC-*, *SO-*, and *SH*-Graph is to provide a systematic representation of service composition patterns along with standard policies that describe the ordering of flow of message among different web services coming from various heterogeneous clouds.

3.1.1 SC-Graph: Service Choreography Graph

Service choreography describes the exchanges of public message, rules of interaction, and the agreements between the participating services at the initial phase of service composition. It is the decentralized perspective that specifies how the individual services interact with each other. A Service Choreography (*SC*-Graph) graph is proposed in Fig. 2 that helps to choreograph the cloud service belongs to the range of consideration of degree of heterogeneity. The *SC*-Graph is defined as a set of heterogeneous service and their interconnection exists between them.

3.1.2 SO-Graph: Service Orchestration Graph

Service Orchestration (*SO*-Graph) helps to orchestras the heterogeneous services lying between the substantial and considerable heterogeneity value of range. The scheduling agent acts as an orchestrator that can retrieve the quality factor of the services as per priority assigned by the scheduling agent. The *SO*-Graph is shown in Fig. 3.

Fig. 2 Service
choreography graph

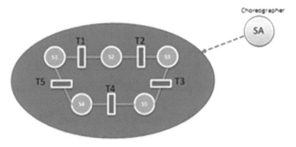

Fig. 3 Service
choreography graph

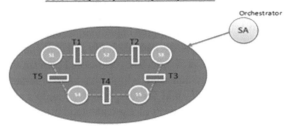

3.1.3 SH-Graph: Service Hybrid Graph

The hybrid-based service composition pattern is a dynamic approach that composes
the service lying between the moderate ranges of degree of heterogeneity. Figure 4
shows the Service Hybrid (*SH*-Graph) where the scheduling agent acts as a controller in the entire composition process. Table 1 shows the corresponding service
composition graph constructs.

Fig. 4 Service hybrid graph

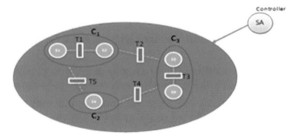

Table 1 Service composition graph constructs

Graph _ID	Graph construct	Meaning
1.		Flow of services
2.		Service "a"
3.		Exclusive-OR flow indicates that after service "a" the direction of flow will go to either "b" or "c" based on certain cases
4.		The parallel flow direction indicates that prior to service "a" the flow from service "b" and "c" will occur in parallel
5.		Message transition
6.		Sequence flow from service "a" to "b"
7.		Cloud "C1"

The salient features of the graph are as follows:

- Solid arrow indicates flow of events.
- Circle represents events/ service.
- Oval shape indicates clouds.

4   Proposed Enterprise Service Composition Petri Net (ESCP)

Enterprise Service Composition Petri nets (*ESCP*) is a graphical modeling tool that models the service composition patterns dynamically and analyzes the behavioral properties of *ESCP* model. The proposed model provides better semantic correctness and effectiveness of composition process in cloud environment. The advantage of using colored Petri nets in designing and analyzing service composition process is to provide a compact and formal way of modeling composition patterns for complex systems, which makes CP nets an effective modeling tool for analyzing enterprise application systems.

4.1   Definition: Enterprise Service Composition Petri Net (ESCP)

An Enterprise Service Composition Petri net (*ESCP*) is described as a bipartite graph which contains several nodes, namely, service_places and service_transitions. The nodes are connected by arcs that represent the state of a service_transitions of the given composition node. Formally, Enterprise Service Composition Petri net (*ESCP*) is described using the conceptual definition of *MAS*-based *ECBS* as discussed in [4].

(a) $ESCP = [P_{sc}, T_{sc}, \Pi_{sc}, In_{sc}, Out_{sc}, H_{sc}, IM_{sc}, W_{sc}]$;

(b) $P_{sc} = \{P_{sc1}, P_{sc2} ..., P_{scm}\}$ is a set of finite places present in the system.

(c) $T_{sc} = \{T_{sc1}, T_{sc2}, T_{scn}\}$ is a set of finite transitions, $S_{sc} \cap T_{sc} = \emptyset$.

(d) $In_{sc}, Out_{sc}, H_{sc}: T_{sc} \rightarrow N_{sc}$ ($N_{sc} = S_{sc} \cup T_{sc}$) are the input, output, and inhibition functions of *ESCP*.

(e) IM_{sc} is the initial marking, where $IM_{sc}: S_{sc} \rightarrow IN_{sc}$ is the initial marking.

(f) $\Pi_{sc}: T_{sc} \rightarrow IN_{sc}$ is the function based on priority that is composed of lower priorities in connection to timed transitions and higher priorities to immediate transitions.

(g) W_{sc} is a weight function $W_{sc}: F_{sc} \rightarrow \{1, 2 ...\}$. $W_{sc}: T_{sc} \rightarrow R_{sc}$ is a function that mapped with a real value to the transitions.

(h) *CF* is the color function of the service composition net. Formally, $CF \rightarrow CF_{sc}$. The color function *CF* mapped to each place P_{sc} to a type CF_{sc}. CF_{sc} defines the color function for service composition that contains different tokens of various colors. $CF_{sc} = CF_s \cup CF_{re}$, CF_s is the color function for service token, and CF_{re} is the color function for request token.

(i) The guard function for service composition pattern G_{sc} mapped to each transition, T_{sc} into a Boolean statement where all variables have such types belongs to ESCP net. The arc expression function Ar_{sc} maps each arc "a" in the node function into an expression of type $CF(P_{sc})$. This means that each arc expression must evaluate to multi-set over the type of the adjacent place, P_{sc}. The initialization function I_{sc} maps each place, P_{sc}, into net expression, which must be of type $CF(P_{sc})$.

4.2 Analysis of Behavioral Properties of ESCP Model

Enterprise Service Composition Petri net (*ESCP*) is an efficient tool to analyze the behavioral facets of service composition process of MAS-based inter-cloud system. The following are some of the crucial behavioral properties of several service composition patterns that have been analyzed using the ESCP model.

The properties of ESCP nets are discussed as follows:

(a) **Safeness**: Using *ESCP* model, the set of places $[P_0...P_n]$ represents a combination of 0 (no token) and 1 (token), which signifies that at each transitions, there will be a token at each place from which the transitions occur, otherwise no token is found. Thus, each place of service composition patterns has a maximum number of token count 1 or 0 and is consider as safe. Since, all the places in the service composition Petri nets using *ESCP* model are safe then the net as a whole can be declared safe.

(b) **Boundedness**: This behavioral properties specify the maximum and minimum number of tokens a place can hold considering all reachable markings in the net. The higher bound for a place in the net signifies the maximal number of tokens a place can hold from any reachable marking. The lower bound for a place in the net signifies the minimal number of tokens a place can hold from any reachable marking. When the value of the upper and lower bounds are equal, the place holds a constant number of tokens and thus it is declared as bound. Since there is no deadlock at any stage within $[P_0...P_n]$, it is bounded.

(c) **Reachability**: Reachability property states that from a given initial marking M_0, each set comprises all the markings that are reached from M_0 through the respective firing of several transitions. In the proposed net, the reachability property holds, if the process starts with initial marking $M_0 = [0\,0]$ of P_0 place and finally reaches to state $M_n = [0\,1]$ of P_n state. Thus, the service composition process for the current request is completed and can achieve the desired goal.

(d) **Dead Transitions**: The term dead transitions are defined as the transitions of a process, which will never be enabled. In all the three types of service composition patterns described in this paper, no dead transitions exist initially, so the composition process for participating service is designed well.

(e) **Liveness**: It is defined as the set of binding elements that remain active through-
    out the process to be complete. In this work, for each service composition pat-
    terns using ESCP model, the initialization of the process sets from P_0 transitions
    T_0 through T_n is fired and place P_n is reached. The ESCP net is continuous and
    hence the liveness property is verified. Therefore, the proposed ESCP is con-
    sidered to live.
(f) **Fairness**: Fairness property is a crucial phenomenon for studying behavioral
    properties of service composition process. No dead activity will be executed in
    the net. The fairness range lies between 0 and 1. Here, we have discussed the
    bounded fairness of the *ESCP* net. An *ESCP* net is said to be Fair net where
    every pair of transitions in the net is in a fair relation.

4.3 Simulation of ESCP Net

The service composition patterns, namely, service choreography, service orchestration, and service hybrid are implemented on online banking system using ESCP net.
The CPN tool is used for the simulation and the corresponding results before and
after transitions are shown. Here, we have considered online banking service system.
This service is considered as complex service as it is composed of several atomic
processes. The online banking service is able to give facilities like mobile banking,
net banking, Demat account, and loan. A client can use mobile banking and net
banking through their mobile number and user_id, password, respectively. Assume
that a client is new client and wants to create an account in a particular bank. The
client specifies the necessary input and the requested output as a query.

   *Inputs: user_id, password, cust_name, mobile_no, address, acc_no, email_id,
amount, bank_info.*

   Outputs: acc_no, amount, card_no, card_type, transaction_id, transaction_type.

   Figures 5, 6, and 7 show the simulation result for service choreography, orchestration, and hybrid patterns using *ESCP* net. In each of the pattern, the Scheduling
Agent (*SA*) fetches the service from the meta-service repository called Hierarchical
Universal Description Discovery and Integration *(HUDDI)* at P_0 place, the relevant
transition T_0 is fired and place P1 is reached. Thus, the process continues till it
reached all the markings (P_n) in the net. Hence, the tokens are allocated and released
dynamically in the given place.

4.4 State-Space Analysis of ESCP

The state-space method of *ESCP* nets makes it possible to validate and verify the
functional correctness of service composition patterns as discussed in this paper. The
state-space process enriches the computation of all the reachable marking states and

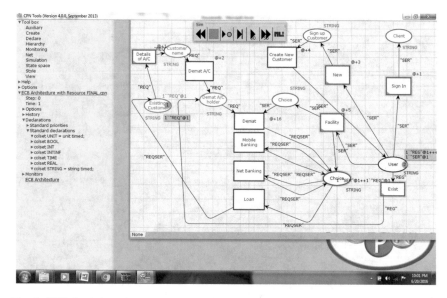

Fig. 5 CPN simulation of service choreography pattern

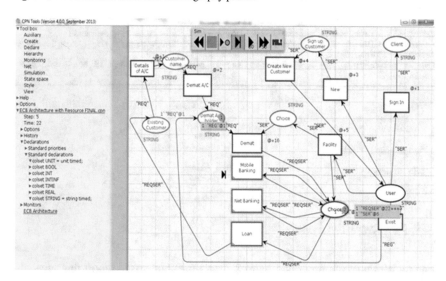

Fig. 6 CPN simulation of service orchestration pattern

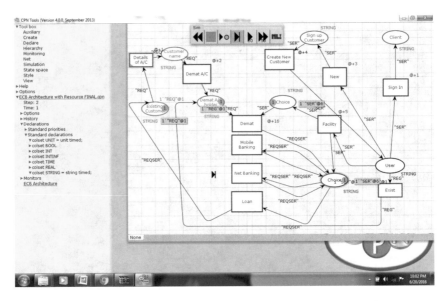

Fig. 7 CPN simulation of service hybrid pattern

```
CPN Tools state space report for:   CPN Tools state space report for:
/cygdrive/C/Users/User/Desktop/final/cygdrive/C/Users/User/Desktop/final year p
Resource FINAL.cpn                  Resource FINAL.cpn
Report generated: Tue Apr 19 00:49:Report generated: Tue Apr 19 00:49:12 2018
 Statistics                          Statistics
-----------------------------------------------------------------------------

    State Space                          State Space
        Nodes:  8                            Nodes:  8
        Arcs:   7                            Arcs:   7
        Secs:   0                            Secs:   0
        Status: Full                         Status: Full

    Scc Graph                            Scc Graph
        Nodes:  8                            Nodes:  8
        Arcs:   7                            Arcs:   7
        Secs:   0                            Secs:   0
```

Fig. 8 State-space report

state transition of the process. Figure 8 shows the state-space report that gives the detailed state-space statistics of the simulation of the model.

5 Conclusion

This work mainly focuses on to formalize the service composition patterns in multi-cloud architecture. The paper also focuses on the analysis and verification of the dynamic properties of service composition models. The proposed Petri net model helps to enrich the service composition dynamics such as accuracy of composing the participant service that leads to the improvement of system performance, design flexibility, performance, and scalability of the composite service and their inter-connections. Further, through *ESCP* concepts, several dynamic properties of like, reachability, safeness, deadlock, etc. are analyzed and verified using online banking service system. The future research directions of the proposed model can be implemented using high-level Petri net concepts that are suitable toward analysis for dynamics of multi-cloud system.

References

1. Khan, G., Sengupta, S., Sarkar, A., Debnath, N.C.: Modeling of inter-cloud architecture using UML 2.0: multi-agent abstraction based approach. In: 23rd International Conference on Software Engineering and Data Engineering (SEDE), pp. 149–154 (2014)
2. Khan, G., Sengupta, S., Sarkar, A., Debnath, N.C.: Web service discovery in enterprise cloud bus framework: T vector based model. In: 13th IEEE International Conference on Industrial Informatics (INDIN), pp. 1672–1677 (2015)
3. Khan, G., Sengupta, S., Sarkar, A.: Priority based service scheduling in enterprise cloud bus architecture. In: IEEE International Conference on Foundations and Frontiers in Computer, Communication and Electrical Engineering (C2E2 2016), SKFGI, Mankundu, India, pp. 363–368 (2016)
4. Khan, G., Sengupta, S. Sarkar, A.: Behavioral modeling of enterprise cloud bus system: high level petri net based approach. In: (SERSC), International Journal of Software Engineering and Its Applications (IJSEIA), Vol. 11, no. 7 (2017), pp. 13–30 (2017)
5. Huhns, M.N., Singh, M.P.: Service-oriented computing: key concepts and principles. IEEE Internet Comput. 9(1), 75–81 (2005)
6. Arsanjani, A.: Service oriented modeling and architecture. IBM Developer Works, pp. 1–15 (2004)
7. Alexandros, K., Aggelos, G., Vassilis, S., Lampros, K., Magdalinos, P., Antoniou, E., Politopoulou, Z.: A cloud-based farm management system: architecture and implementation. J. Comput. Electron. Agric. 100, 168–179 (2014)
8. Chang, S.H., La, H.J., Bae, J.S., Jeon, W.Y., Kim, S.D.: Design of a dynamic composition handler for esb-based services. In: IEEE International Conference on e-Business Engineering, (ICEBE 2007), pp. 287–294 (2007)
9. Dubey, S., Agrawal, S.: QoS driven task scheduling in cloud computing. Int. J. Comput. Appl. Technol. Res. 2(5), 595–meta (2013)
10. Kuzu, M., Cicekli, N.K.: Dynamic planning approach to automated web service composition. Appl. Intell. 36(1), 1–28 (2012)
11. Gutierrez-Garcia, J.O., Sim, K.M.: Self-organizing agents for service composition in cloud computing. In: IEEE Second International Conference on Cloud Computing Technology and Science (CloudCom), pp. 59–66 (2010)
12. Ghanbari, S., Othman, M.: A priority based job scheduling algorithm in cloud computing. Procedia Eng. 50, 778–785 (2012)
13. Azgomi, M.A., Entezari-Maleki, R.: Task scheduling modelling and reliability evaluation of grid services using coloured Petri nets. Future Gen. Comput. Syst. 26(8), 1141–1150 (2010)
14. Hu, H., Li, Z.: Modeling and scheduling for manufacturing grid workflows using timed Petri nets. Int. J. Adv. Manuf. Technol. 42(5–6), 553–568 (2009)

15. Yang, Y., Tan, Q., Xiao, Y.: Verifying web services composition based on hierarchical colored petri nets. In: Proceedings of the First International Workshop on Interoperability of Heterogeneous Information Systems, pp. 47–54 (2005)
16. Jun, Z., Ngan, W.H., Junfeng, L., Jie, W., Xiaoming, Y.: Colored petri nets modeling of multi-agent system for energy management in distributed renewable energy generation system. In: Asia-Pacific Power and Energy Engineering Conference (APPEEC), pp. 28–31 (2010)

Part III
VLSI and Graph Algorithms

Generation of Simple, Connected, Non-isomorphic Random Graphs

Maumita Chakraborty, Sumon Chowdhury and Rajat Kumar Pal

Abstract In graph theory, generation of random graphs finds a wide range of applications in different scheduling problems, approximation algorithms, problems involving modeling and simulation, different database applications, and obviously to test the performance of any algorithm. The algorithm, which has been devised in this paper, is mainly for the purpose of providing test bed for checking performance of other algorithms. It generates different non-isomorphic graph instances of a given order and having unique number of edges. The number of such instances possible for a graph of given order has also been subsequently formulated. Different such graph instances of different orders, generated in a uniform computing environment, and the computing time required for such generations have also been included in this paper. The simplicity and efficiency of the algorithm, subsequently proved in the paper, give us a new insight in the area of random graph generation and have called for further research scope in the domain.

Keywords Random graph · Non-isomorphic graph · Connected graph · Graph generation

M. Chakraborty (✉)
Department of Information Technology, Institute of Engineering and Management, Salt Lake, Sector V, Kolkata 700091, India
e-mail: maumita.chakraborty@gmail.com

S. Chowdhury
Tata Consultancy Services, Gitanjali Park, Action Area II, New Town, Kolkata 700156, India
e-mail: c.sumon19@gmail.com

R. K. Pal
Department of Computer Science and Engineering, University of Calcutta, JD-2, Sector III, Salt Lake, Kolkata 700106, India
e-mail: pal.rajatk@gmail.com

© Springer Nature Singapore Pte Ltd. 2020
R. Chaki et al. (eds.), *Advanced Computing and Systems for Security*,
Advances in Intelligent Systems and Computing 995,
https://doi.org/10.1007/978-981-13-8962-7_6

69

1 Introduction

The study of random graphs in their own right began earnestly with the seminal paper of Erdos and Renyi [6] many decades ago. A random graph is a graph in which properties such as the number of graph vertices, graph edges, and connections between them are determined in some random way [9, 14]. Random graphs generated from any graph generator algorithm are very useful in providing a test bed for scientists to work with, while testing performance of approximation algorithms. Varied sets of graphs not only say how well an algorithm performs but are also used to determine some of the worst-case solutions to different algorithms. Random graph generation may also be used in several other domains like scheduling problem for distributed/parallel systems, several database applications, modeling and simulating many complex real-world systems, uniform sampling of random graphs, and many more. Real-world networks are often compared to random graphs to assess whether their topological structure could be a result of random processes.

To cater to some of the needs mentioned above, we have developed a simple but efficient method of generating random graph instances. The main intention behind development of the algorithm is to provide a test bed for some algorithms which demanded for generation of only non-isomorphic graphs [5, 13]. That is why we focused only on the generation of some random non-isomorphic graph instances. It has been found out that the order of the graph being given we can eventually formulate the number of such instances possible for a specified range of edges.

2 Literature Survey

A number of academicians from all over the world have come up with different algorithms for the random graph generation problem for the last few years [2–4, 6, 7, 9–12]. The input and output set as well as the purpose behind such generation may differ for different algorithms. In this section, we have considered few such algorithms and have described, very briefly, about the input and output sets as well as the target application areas.

- The algorithm by Horn et al. [7] has taken into consideration the number of nodes in the graph, the number of levels of the graph, the degree for each node, and the height and width of each graph. The development of this graph generator was mainly to aid in testing the design and development of an approximation algorithm. Hence, here the graph outputs need to be acyclic, directed, and rooted.
- The algorithms by Viger and Latapy [10, 11] claims to be very suitable for practical use because of its ability to generate very large graphs, which increases its efficiency, as well as the ease to implement it, which proves its simplicity. The large graphs are supposed to have more than one million vertices, as often needed in simulations.

- The algorithm by Bayati et al. [2] provides more insight into random graphs with a given expected degree sequence. It has also used similar ideas to generate random graphs with large girth, which are used to design high-performance Low-Density Parity-Check (LDPC) codes, by generating random bipartite graph instances with an optimized degree sequence.
- The algorithm by Cordeiro et al. [4] provides a standard implementation of classical task graph generation methods, as well as an in-depth analysis of each generation method, used in the scheduling algorithms.
- Several database applications, generation of random structures for simulation, sampling, etc. called for the proposal of a data parallel algorithm for random graph generation by Nobari et al. [9]. Data parallel algorithms focus on distributing the data across different nodes, which operate on the data in parallel. The authors claim that their algorithms are efficient for generating graphs that fit into main memory for further processing and also for larger graphs to be further processed concurrently, in parallel, or in a streaming fashion.
- The main aim of the algorithm by Wang et al. [12] is to generate clustered random graphs in large scale which are suitable for real network. The algorithm also claims to have advantages in randomness evaluation and computation efficiency when compared with the existing algorithms.
- Bhuiyan et al. [3] have presented an efficient parallel algorithm for generating random graphs with given degree sequence. They have also claimed to estimate the number of possible graphs with a given degree sequence, which can be helpful in studying various structural properties and dynamics over a network.

3   Our Proposed Algorithm, *CreateNonIsomorphicGraphs*

This paper is mainly targeted toward the development of a simple but efficient algorithm to obtain different graph instances of a given order. While studying and implementing different algorithms for generating all possible spanning trees [15] of a simple and connected graph, we felt the need for a set of graph instances, with varied order and density. Now, we know that isomorphic graphs [5, 13] do have same number of vertices and edges, along with their one-to-one relation in mapping. Thus, it is a fact that isomorphic graphs will be generating same number and same set of spanning trees. As a result, the input set for all spanning tree generation problem should be ideally a set of mutually non-isomorphic graph structures, which will provide more insight into how the number of generated spanning trees and the time taken for this generation vary with varying number of vertices and edges of the graphs.

While generating all possible non-isomorphic graphs [5, 13] with a given order, we found that the output ranges from a complete graph (with a maximum possible number of edges, which is undoubtedly $n(n-1)/2$ for a graph of order n) to a spanning tree (with minimum possible edges, which is $n-1$ for a graph of order n) [5]. Thus, if S be the set of all simple, connected, and non-isomorphic graphs with n vertices and

m edges, then $S = \{G = (n, m) | n - 1 \leq m \leq n(n - 1)/2\}$, where G is the set of all graphs with n vertices and m edges.

Another significant observation in this regard is that every non-isomorphic graph is a subset of the complete graph of a particular order. Thus, S can be generated from the complete graph of n vertices by systematic deletion of edges. Since isomorphism checking takes exponential time, generation of graphs with the same value of m has been avoided while designing the algorithm under consideration. Let S' be a subset of S, where each graph has a unique value of m. Maximum cardinality of S' is

$$\left.|S'|\right._{max} = (n(n - 1)/2) - (n - 1) + 1 = (n^2 - 3n + 4)/2.$$

Procedure *CreateNonIsomorphicGraphs(n)*, shown in Fig. 1, generates S' with cardinality $\left.|S'|\right._{max}$ by taking n as input. It initially creates a complete graph, K_n, and then generates the rest of the $(n^2 - 3n + 4)/2 - 1$ graphs by deleting one edge at a time from K_n. The algorithm also ensures that no disconnected graphs are being generated, by taking into account the minimum degree of vertices (d_{min}), after removal of each edge from the graph. The total number of edges that can be removed from a vertex should not be less than d_{min}; otherwise, it may lead to disconnected component(s). In our algorithm, we have not considered the disconnected graphs generated as we were interested in generating only spanning trees from a given graph, instead of generating spanning forests [5, 15].

The algorithm at a glance for *CreateNonIsomorphicGraphs* has been given in Fig. 1 for better understanding. The algorithm, as shown in Fig. 1, makes use of certain variables like i, tt, F, f, and x for computation and storing intermediate results. Moreover, it is worth mentioning that the above procedure can be modified to generate graphs with any specified range of edges.

As mentioned earlier, the different graph generation algorithms differ not only by their application areas but also by their input and output sets. Now the question of comparing the performances of two or more algorithms come only when their input and output sets are same. Our algorithm is very much unique in its output set compared to other algorithms discussed in the paper, and hence its performance is not really comparable with the other ones. We can still claim that our method of graph generation is unique and robust. Novelty of our approach lies in computing only non-isomorphic graphs of order n where some k edges have been deleted, where $1 \leq k \leq (n - 1)(n - 2)/2$. Other existing methods may add edges (starting from n isolated vertices) or employ some other tricks.

4 Data Structures and Complexity Issues

In *CreateNonIsomorphicGraphs*, a complete graph is being formed and stored in the form of adjacency matrix, whose storage requirement is $O(n^2)$, which can be further reduced to $O(n+m)$ using adjacency list representation of graphs [1, 8]. From this

```
proc CreateNonIsomorphicGraphs(n)
Begin
Let K_n be the complete graph with n vertices
tt ← K_n
Print tt
Let d_min be the minimum degree of tt
F ← Φ
x ← 0
For each i ∈ {1, 2, ..., d_min} do
   e ← any edge selected randomly from tt
   tt ← tt - e
   tt ← tt ∪ F
   Print tt
   tt ← tt - F
   x ← x + 1
Endfor
If x > (n²-3n+4)/2 then
   Go to Step 27
Else
   d_min ← minimum degree of tt
   If d_min > 1, then
   Go to Step 7
   Else
      F ← the edge incident on the minimum degree vertex
      F ← F ∪ f
      tt ← tt - f
      d_min ← minimum degree of tt
   Endif
Endif
Endproc
```

Fig. 1 Algorithm for generating a set of simple connected non-isomorphic graphs with n vertices

complete graph, we are gradually generating the different random non-isomorphic graphs, by just deleting edges one by one. Hence, no extra space is consumed for them.

Formation of initial complete graph of order n takes $O(n^2)$ time. To select a random edge and deleting the corresponding entries from the adjacency matrix, in order to generate another graph, take constant time. This process of deleting edges is repeated at most for $(n-1)$ times. Hence, the time taken for generation of all graphs, other than the complete graph K_n, is $O(n)$. As a result, the total time complexity for *CreateNonIsomorphicGraphs* comes out to be $O(n^2)$.

5 A Worked Out Example

In this section, we explain the working procedure of *CreateNonIsomorphicGraphs*
with an example. Figure 2 shows the series of non-isomorphic random graphs that
have been generated following the steps of *CreateNonIsomorphicGraphs*, starting
from a complete graph of order 5, i.e., K_5, as in Fig. 2a.

The degrees of all vertices of K_5, in Fig. 2a, are same and obviously greater than
one. Hence, we can choose any vertex of K_5 and remove an edge from it. If any
random edge $\{v_3, v_4\}$ is removed, the graph generated is shown in Fig. 2b. Now the
vertices v_3 and v_4 are having the same minimum degree, which is three (still greater
than one). Hence, selecting the edge $\{v_2, v_3\}$ randomly gives us the graph in Fig. 2c.
Now, d_{min} becomes two (for vertex v_3). Hence, there is still a provision to generate
more graphs from the graph of Fig. 2c. If the random edge deleted this time be $\{v_2,$
$v_4\}$, then the resulting graph is what is shown in Fig. 2d. It is evident from the graph
in Fig. 2d that $d_{min} = 2$. Hence, another random edge can be deleted from it, say $\{v_4,$
$v_5\}$.

The resultant graph has been shown in Fig. 2e. The graph in Fig. 2(e) clearly
shows that $d_{min} = 1$. Hence, there is a chance that if in the next iteration the edge $\{v_1,$
$v_4\}$ gets selected randomly, then it will generate a disconnected graph, which is not
acceptable. On the contrary, there is also a chance that some other edges from some
other vertex get selected, after whose removal the graph still remains connected. To
avoid checking the connectedness of a graph, we have not considered any further

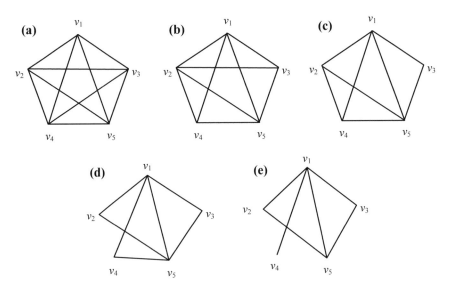

Fig. 2 a Complete graph of order 5, K_5, **b** Graph with one edge removed from K_5, **c** Graph with
two edges removed from K_5, **d** Graph with three edges removed from K_5, **e** Graph with four edges
removed from K_5

instances, as already discussed in earlier sections. Thus, out of the seven possible graphs of order five that can be generated, only five, which are guaranteed to be connected, have been considered in this example.

It is worth mentioning that the sequence of edges that have been selected for removal in each step could have been different, as a result of which some other random graphs could have been generated. The number of random graphs also could have been different. This process works well for graphs of any order.

6 Experimental Results

The algorithm *CreateNonIsomorphicGraphs*, developed in this paper, has been implemented on an Intel Core $i3$ quad-core processor running at 2.4 GHz, with 6 GB RAM. Random graph instances have been generated for graphs of order ranging from $|V| = 10\,to\,1000$, $|V|$ being the total number of vertices (i.e., n). The CPU time, taken by the algorithm to execute, for each such generation is shown in seconds in Table 1. As discussed in the earlier section, some graph instances may become disconnected during the generation process. Hence, for each such generation, we have shown the number of possible non-isomorphic graphs (*PG*) that can be generated by our algorithm, the number of valid/connected non-isomorphic graph instances (*VG*), as well as the number of invalid/disconnected instances (*PG–VG*) generated during an execution. We have also computed the percentage of valid graphs over possible number of graphs, as shown in Table 1. From Table 1, we can readily conclude the following, proving the efficiency of our algorithm:

- The graph generation has been done very fast for very high-ordered graphs.
- The number of disconnected instances is much low compared to the connected instances (while deleting n edges one after another, at random), which is evident from the parameter (*VG/PG* $\times$ 100) in the last column of the table. Parameter (*VG/PG* $\times$ 100) is also noted to get better with increase in the order of the graphs.

7 Conclusion

Random graph generation has been a topic of interest for computer scientists for different purposes. This generation process is often required for testing performances of various other algorithms also. We have a similar intention while developing our random graph generation algorithm, *CreateNonIsomorphicGraphs*, which has been designed to generate only a set of simple, connected, and non-isomorphic graphs of any order. In the result section of this paper, we have generated such graph instances of order ranging from 10 to 1000. The efficiency of our algorithm has been proved by the very little time it takes to generate all desired graphs of high order, as well as

Table 1 Experimental results of generating random non-isomorphic graphs of order 10 to 1000

Graph order (n)	Number of possible non-isomorphic graphs that can be generated (PG)	Number of valid/connected non-isomorphic graphs generated (VG)	Number of invalid/ disconnected graphs generated (PG–VG)	CPU time (secs)	Percentage of valid graphs to total possible graphs (VG/PG × 100)
10	37	28	9	0	75.7
20	172	142	30	0	82.6
30	407	368	39	0	90.4
40	742	665	77	0	89.6
50	1177	1055	122	0	89.6
60	1712	1585	127	0	92.6
70	2347	2202	145	0	93.8
80	3082	2873	209	0	93.2
90	3917	3582	335	0	91.4
100	4852	4561	291	0	94
200	19702	18987	715	0	96.4
300	44552	43355	1197	1	97.3
400	79402	76740	2662	2	96.6
500	124252	121930	2322	3	98.1
600	179102	176289	2813	6	98.4
700	243952	240132	3820	7	98.4
800	318802	315278	3524	13	98.9
900	403652	399708	3944	19	99
1000	498502	493268	5234	26	99

the high percentage of connected instances being generated, which even gets better with increase in order of the graphs.

In our paper, we have claimed to generate only a set of non-isomorphic graphs with unique size, m (i.e., the number of edges). There may be many more non-isomorphic graphs within the possible range of edges, which have not been generated here considering the fact that isomorphism checking takes exponential time. Moreover, our requirement is highly satisfied with what we have generated. Hence, we have decided to leave this as an open problem for future researchers. Our algorithm is flexible enough to be modified easily to generate all possible non-isomorphic graphs within any specified range of edges. Thus, it provides the basic guidelines and shows a new direction to future researchers to come up with many newer, as well as enriched ideas in this domain.

References

1. Aho, A.V., Hopcroft, J.E., Ullman, J.D.: Data Structures and Algorithms. First Edition. Pearson (1983)
2. Bayati, M., Kim, J.H., Saberi, A.: A sequential algorithm for generating random graphs. Algorithmica (Springer) **58**(4), 860–910 (2010)
3. Bhuiyan, H., Khan, M., Marathe, M.: A parallel algorithm for generating a random graph with a prescribed degree sequence. In: arXiv preprint: 1708.07290 (2017)
4. Cordeiro, D., Mounie, G., Perarnau, S., Trystram, D., Vincent, J.M., Wagner, F.: Random graph generation for scheduling simulations. In: Proceedings of Third International Conference on Simulation Tools and Techniques (SIMUTools'10), Article No. 60 (2010)
5. Deo, N.: Graph Theory with Applications to Engineering and Computer Science. Prentice Hall of India Pvt. Ltd., New Delhi (2003)
6. Erdos, P., Renyi, A.: On the evolution of random graphs. Publ. Math. Inst. Hung. Acad. Sci. **5**, 17–61 (1960)
7. Horn, M.V., Richter, A., Lopez, D.: A random graph generator. In: Proceedings of 36th Annual Midwest Instruction and Computing Symposium, Duluth, Minnesota (2003)
8. Horowitz, E., Sahni, S., Anderson, S.: Fundamentals of Data Structures in C, 2nd edn. Universities Press Pvt. Ltd., Hyderabad, India (2008)
9. Nobari, S., Lu, X., Karras, P., Bressan, S.: Fast random graph generation. In: Proceedings of the 14th International Conference on Extending Database Technology, Uppsala, Sweden, pp. 331–342 (2011). https://doi.org/10.1145/1951365.1951406
10. Viger, F., Latapy, M.: Efficient and simple generation of random simple connected graphs with prescribed degree sequence. In: Computing and Combinatorics, COCOON 2005. Lecture Notes in Computer Science (Springer, Berlin, Heidelberg) 3595, pp. 440–449 (2005). https://doi.org/10.1007/11533719_4
11. Viger, F., Latapy, M.: Efficient and simple generation of random simple connected graphs with prescribed degree sequence. J. Complex Netw. **4**(1), 15–37 (2016)
12. Wang, C., Lizardo, O., Hachen, D.: Algorithms for generating large-scale clustered random graphs. Network Sci. **2**, 403–415 (2014)
13. Weisstein, E.W.: (IG) isomorphic graphs. In: MathWorld—A Wolfram Web Resource. http://mathworld.wolfram.com/IsomorphicGraphs.html. Accessed 18 Jan 2019
14. Weisstein, E.W.: (RG) random graph. In: MathWorld—A Wolfram Web Resource. http://mathworld.wolfram.com/RandomGraph.html. Accessed 18 Jan 2019
15. Wikipedia (ST) Spanning Tree. https://en.wikipedia.org/wiki/Spanning_tree. Accessed 18 Jan 2019

Bottleneck Crosstalk Minimization in Three-Layer Channel Routing

Tarak Nath Mandal, Kaushik Dey, Ankita Dutta Banik, Ranjan Mehera and Rajat Kumar Pal

Abstract Channel routing and crosstalk minimization are important concerns while we talk about high-performance circuits for two-, three-, and multilayer VLSI physical design automation. Interconnection among the net terminals satisfying constraints in an intelligent way is a necessity to realize a circuit within a minimum possible area, as this is a principal requirement to diminish cost as well as to augment yield. Introduction of a layer of interconnects may increase the cost of routing; however, as the area is minimized, cost is reduced as well. Eventually, the total wire length is also reduced. Therefore, there are several trade-offs. Besides, in high-performance routing, a designer is supposed to lessen the amount of electrical hazards, viz., crosstalk as much as possible. In this paper, we particularly work on minimizing bottleneck crosstalk in the three-layer HVH routing model. Here, along with area minimization, computed circuits' performance has also been enhanced by computing precise bottleneck crosstalk HVH channel routing solutions. By the way, the specified crosstalk minimization problem is NP-hard. Thus, in this paper, heuristic algorithms have been devised for computing optimized bottleneck crosstalk channel routing solutions. Computed results of our proposed algorithms are greatly encouraging.

Keywords Bottleneck crosstalk · Channel routing problem · Crosstalk minimization · Manhattan routing model · NP-hard · Three-layer routing

T. N. Mandal · K. Dey · A. D. Banik · R. K. Pal (✉)
Department of Computer Science and Engineering, University of Calcutta, JD-2, Sector – III, 700106 Saltlake, Kolkata, India
e-mail: pal.rajatk@gmail.com

T. N. Mandal
e-mail: tarak.sap2016@gmail.com

K. Dey
e-mail: kaushikdey59@gmail.com

A. D. Banik
e-mail: duttabanik.ankita@gmail.com

R. Mehera
Subex Inc, 12303 Airport Way, Suite 390, Broomfield 80021, CO, USA
e-mail: ranjan.mehera@gmail.com

© Springer Nature Singapore Pte Ltd. 2020
R. Chaki et al. (eds.), *Advanced Computing and Systems for Security*,
Advances in Intelligent Systems and Computing 995,
https://doi.org/10.1007/978-981-13-8962-7_7

1 Introduction

Channel routing is a distinct and significant problem in VLSI physical design automation [1, 2]. In this problem, it is required to generate a specified wiring of a set of terminals present on the periphery of different modules on two opposite sides of a rectangular routing region, primarily using minimum possible area. Incidentally, channel routing is one of the most crucial existing detailed routing schemes [2, 3].

A *channel* is a rectangular local routing region that has two open ends, the left and right sides of the rectangle. The other two sides, viz., the upper and lower sides of the rectangle, have two rows of fixed terminals. In the grid-based routing model, the terminals are assumed to be aligned vertically in *columns* that are usually equispaced along the length of the channel. A set of terminals that need to be electrically connected together is called a *net*. In Fig. 1, two columns having the same number (other than zero) uniquely define a *two-terminal net*. In general, a channel contains multiterminal nets as well. Zeros are non-terminals, not to be connected.

In rectilinear wiring, a *vertical (net) segment* is a wire that lies in a column, whereas a *horizontal (net) segment* is a wire that lies in a track. *Tracks* are horizontal (grid) lines that are usually equispaced along the height of the channel, parallel to the two rows of (fixed) terminals.

A *route* for a net is a collection of horizontal and vertical segments spread across different layers connecting all the terminals of the net. A *legal wiring* of a channel is a set of routes that satisfy all the prespecified conditions, where no two net segments used to connect different nets overlap (or short-circuited, even at a point) on the same layer of interconnection. A (complete) legal wiring of a given channel instance is also called a *feasible routing solution* in some specified channel routing model.

The *channel routing problem* (CRP) is stated by two m element vectors *TOP* and *BOTTOM*, and a number t; the usual objective is to find a feasible routing solution for the channel using no more than t tracks, if one exists. An instance (not a solution) of the CRP is shown in Fig. 1, where we have an assignment of intervals of the nets present in this channel to four tracks only. Let L_i (R_i) be the leftmost (rightmost) column position of net n_i, then $I_i = (L_i, R_i)$ is known as the *interval* (or *span*) of the net.

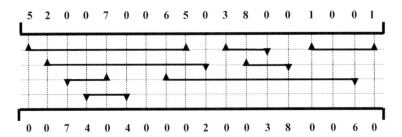

Fig. 1 An example channel of eight nets. Arrows indicate that the terminals to be connected, either on top or on bottom, to complete all interconnections of the nets present in the channel

In this work, we have assumed the crosstalk minimization problem as an important high-performance factor. Due to advancement of fabrication technology, feature size reduces and the devices are placed in closer proximity while the circuits are expected to operate in higher frequencies. As a result, electrical hazards, viz., *crosstalk* between net segments are evolved, which is proportional to the capacitive coupling that depends on the coupling length (i.e., the total length of overlap between the segments of two different nets placed nearby). Crosstalk is also proportional to the frequency of operation and inversely proportional to the separating distance between wires. In our work, overlapping between a pair of segments of two different nets is assumed negligible if these are placed in nonadjacent tracks, as crosstalk is significantly reduced.

More crosstalk means more signal delay and deteriorated circuit performance. Thus, it is desirable to develop channel routing algorithms that not only minimize the number of tracks (i.e., channel area) but also crosstalk. Work on routing channels with reduced crosstalk is very important from high-performance criterion for VLSI circuit synthesis [2, 4].

Thus, it is anticipated if a routing solution could be computed where the amount of total crosstalk is mostly reduced, which is an NP-complete problem [5, 6]. However, it has been observed that even if total crosstalk is reduced, performance of the solution may not be achieved due to long overlapping of a pair of nets' horizontal segments assigned to (two) adjacent tracks. Hence, a new problem, specifically the bottleneck crosstalk minimization, comes into the picture, which has also been proved to be NP-complete [5, 6]. As a natural consequence, researchers are intended to devise heuristic algorithms or to employ some soft computing technique such that bottleneck crosstalk is taken care of. One such work has been reported in [7] to design heuristic algorithms for two-layer channel routing. In this paper, several such experimentations have been performed in order to reduce bottleneck crosstalk in computing three-layer channel routing solutions to enhance the circuit performance.

The paper is organized as follows. Section 2 includes some basics relating to channel routing problem and crosstalk minimization. The problems are formulated in Sect. 3. Algorithms have been devised in Sect. 4. Section 5 comprises the experimental results along with necessary discussions. The paper is concluded in Sect. 6.

2 Preliminaries

The channel routing algorithms are developed based on a routing model under consideration. One such most practical and appropriate model of routing that IC industries have assumed in developing and manufacturing much of the VLSI chips is the grid-based reserved layer non-overlap Manhattan channel routing model [2, 3]. The concept of grid-based routing model has already been discussed. In the three-layer reserved layer routing model, HVH, two layers are reserved for horizontal segments of nets and the other layer in between is reserved for vertical net segments only [1, 2]. Connections between two orthogonal segments of the same net are achieved by a via, located at the junction point of the grid. This model is of course a non-overlap routing model, as the horizontal layers are not assumed adjacent (even if wire segments of two different nets are assigned to the same track of these two layers that may overlap).

Finally, the Manhattan routing allows only rectilinear wiring for making the interconnections, in which horizontal net segments are parallel to the two rows of fixed terminals and knock-knees are not allowed [1, 2]. In no-dogleg routing, the horizontal segment of a net is assigned to a track only (and not split into two or more for their assignment to two or more tracks) in a routing solution.

Based on the particular routing model above, the CRP is characterized by two important constraints: horizontal constraints and vertical constraints [2, 3]. The *horizontal constraints* signify if a pair of horizontal segments of two different nets is assignable to the same track (in a horizontal layer), and the *vertical constraints* specify the assignment of net intervals to track along the height of the channel. The graphical representation of these two constraints depicts the *horizontal constraint graph* (*HCG*) and the *vertical constraint graph* (*VCG*), respectively [2, 3]. By the way, the HCG is an interval graph, which is triangulated [8]. On the other hand, the VCG is an arbitrary directed graph. The maximum number of nets passing through a column (in a channel) is known as the *channel density*, which is same as the size of the clique in the HCG, and denoted as d_{max} [2]. In contrast, v_{max} is the length of the longest path in an acyclic VCG [2], and $max(\lceil d_{max}/2 \rceil, v_{max})$ is a lower bound on the number of tracks in the aforementioned three-layer channel routing model [2].

In devising channel routing algorithms, often we view horizontal constraints with the help of *horizontal non-constraint graph* (*HNCG*), which is the complement of HCG [9]. We may note that as an HCG is an interval graph, the corresponding HNCG is a comparability graph and the nets belonging to a clique in the HNCG can securely be assigned to a track (obeying vertical constraints) to compute a routing solution [2, 6].

We define that the amount of crosstalk between the horizontal segments (i.e., intervals) of two different nets assigned to two adjacent tracks in a layer of a given routing solution is same as the amount of overlapping of their horizontal spans. If two intervals do not overlap, there is no horizontal constraint between the nets, i.e., if there is an overlap of a pair of horizontal segments (in a layer), then only there is a scope of having *accountable* crosstalk (other than negligible or no crosstalk) between them.

On the other hand, *bottleneck crosstalk* with respect to a feasible routing solution is the maximum amount of crosstalk (to be defined by us) due to overlapping between any pair of adjacent horizontal segments of two different nets (in a layer). Whether a pair of nets (overlapping to each other) is assignable to two adjacent tracks (in a layer) is the major concern in the problem of bottleneck crosstalk minimization. Hence, the bottleneck crosstalk minimization problem is the problem of computing a practical routing solution with a given number of tracks such that the maximum amount of crosstalk between a pair of (net) intervals in a horizontal layer is within a bound.

In the next section, we briefly formulate the problems under consideration toward devising subsequent algorithms in Sect. 4.

3 Formulation of the Problems

3.1 Computation of Minimum Area Three-Layer Channel Routing Solutions

The input to the CRP is a channel specification that comprises the top vector and the bottom vector of a channel. There are some inherent constraints that have already been stated in the previous section. In this formulation, we represent horizontal constraints with the help of HNCG and vertical constraints with the help of VCG. Note that the VCG should be acyclic in order to compute a feasible solution in the specified three-layer channel routing model as there is only one vertical layer in between [2].

If a channel instance is free from any vertical constraint, a routing solution using $\lceil d_{max}/2 \rceil$ number of tracks is assured in the three-layer HVH routing model [2, 10–12]. However, a general channel specification contains both horizontal and vertical constraints that we are supposed to obey in devising any routing algorithm in the model under consideration toward achieving a desired routing solution.

Note that an HNCG is a (undirected) comparability graph that can be oriented based on the presence of the nets (or intervals) from left to right along the length of the channel. Here an edge connects a pair of vertices if the associated intervals do not overlap in the channel. Now, there are several factors involving a net in a channel that can be sequenced (in the form of tuple weights) based on some priority of assigning them to tracks [2, 12], or that can be assumed in combination as a single weight to each allied net. Next, the maximum weighted clique computation algorithm of a comparability graph [8] is utilized to compute a set of most desirable nets for their assignment to a track under consideration of one horizontal layer [12]. However, if such nets in a clique are vertically constrained to each other, those cannot be assumed at a time (for their assignment to a track). To resolve this issue, we do the following.

It is very clear that each of the nets whose vertices are source vertices in the VCG is assignable in separation to the topmost track of a layer of the channel. However, they might have horizontal constraints among themselves. We may note that all the information relating to horizontal non-constraints are there in the HNCG (in the form of edges). So, if possible, a maximum weighted clique from the HNCG is supposed to be computed such that the corresponding nets (that comprise a non-overlapping set of intervals) would be considered for their assignment to the (current) topmost track in one horizontal layer.

Thus, in our formulation what we do, first we compute the set of source vertices from the VCG, then we extract the induced subgraph from the HNCG comprising the said set of source vertices, which are weighted based on some weight metric defined by us and acquired from the channel itself. Then we compute a maximum weighted clique from this induced HNCG (which is a comparability graph) such that the associated nets in the clique are safely assignable to the (current) topmost track in the first horizontal layer. If at least one vertex of the induced HNCG remains to assign, we compute another maximum weighted clique from the residual induced HNCG such that the associated nets in this clique are securely assignable to the (current) topmost track in the second horizontal layer.

Clearly, the algorithm is iterative in nature. In each iteration, the set of source vertices is differentiated from the VCG for their assignment to the current topmost track of both the horizontal layers. Then the vertices corresponding to the nets assigned to the current topmost track are deleted from the VCG and the HNCG (along with their adjacent edges) to start with the next iteration, until and unless all the nets are assigned to tracks. Afterward, all vertical (net) segments are assigned to connect the respective net terminals to make the routing complete using the only vertical layer which is flanked by horizontal layers.

In addition, the number of such iterations (of the devised heuristic algorithm) is same as the number of tracks required for the channel under consideration. This is how probably a minimum area three-layer HVH routing solution is computed in which all the constraints present in the given channel specification are conformed.

3.2   Computation of Bottleneck Crosstalk Three-Layer Routing Solutions

As total crosstalk minimization in two-layer VH channel routing for general channel specifications by permuting tracks of a given routing solution is an NP-hard problem [4, 5], in time devising a heuristic algorithm may be a natural decision [13, 14]. Subsequently, it has been proved that the bottleneck crosstalk minimization problem is also NP-complete for two- and three-layer channel routing [5, 6]. Recently, an algorithm has been devised to minimize bottleneck crosstalk in two-layer VH channel routing, where area has also been minimized [7]. In this algorithm, initially, it has been attempted to compute a compacted two-layer VH channel routing solution, and then tried to reach a specified bottleneck crosstalk routing solution by moving nets to some other tracks. If in the end a desired bottleneck value is not reached (for a pair of nets assigned to adjacent tracks with *long* overlapping), a blank track is introduced (means the channel area is compromised), if at all necessary, to achieve the stipulated bottleneck value.

For any three-layer HVH feasible routing solution, S of t tracks, another feasible three-layer HVH routing solution, S' with zero crosstalk can be computed by introducing t-1 blank tracks, where each blank track is introduced in between two adjacent tracks in S. However, S' will be a solution with almost $2t$ tracks and so there is a trade-off between the routing area and the resulting crosstalk in routing a channel, and computing a routing solution with more channel area is also not acceptable in general.

In this work, our objective is to develop a heuristic algorithm, which is essentially greedy in nature, by considering the nets in S in a sequence and rerouting them to some other tracks (if possible, to some other horizontal layers as well), such that the new valid routing solution, S' may contain a maximum of either 60%, or 75%, or 90% bottleneck crosstalk in contrast to the maximum amount of crosstalk available in S. This may be accomplished by exchanging a pair of nets (or groups of nets) or by reallocating an interval to a blank space in some other track (or in some other layer) justifying the constraints and crosstalk is also reduced involving the net(s). Following some rational sequence of nets (or groups of nets), intervals are attempted to move for their reassignment to some other tracks such that the bottleneck crosstalk is realized.

In the next section, we introduce the proposed algorithms. Performance of the algorithms is also presented in the subsequent section of experimental results.

4 Proposed Algorithms

4.1 Algorithms for Computing Minimum Area and Minimum Crosstalk Three-Layer Channel Routing Solutions

Computing a three-layer no-dogleg minimum area routing solution in the specified channel routing model is a hard problem for general channel instances [15]. Anyway, as has been formulated in the earlier section, the proposed heuristic algorithm computes the channel density (d_{max}) and the length of the longest path (v_{max}) in the VCG of a channel, if the VCG is acyclic. Then an iterative algorithm is developed as follows where, in each iteration, an induced HNCG is computed among the set of source vertices in the VCG, in which an edge is directed from left to right for each non-overlapping pair of nets along the length of the channel.

Our proposed algorithm computes a composite weight for each of the vertices in the HNCG. Thus, from the induced HNCG (among the set of source vertices, at the beginning of some iteration), two successive maximum weighted cliques are computed so that the nets in the respective cliques are assigned to the current (topmost) track of the two horizontal layers under consideration. Then, the VCG and the HNCG are modified after deleting the vertices (along with their connected edges) for the nets just assigned to a track. Then, the next iteration begins based on the set of source vertices in the updated VCG. This process is continued until all the nets in the channel are assigned to tracks.

When the initial minimum area three-layer HVH routing solution is obtained, we most importantly compute the amount of maximum crosstalk, c_{max}, between a pair of nets assigned to adjacent tracks in S. This is because the bottleneck crosstalk is then defined based on this maximum crosstalk, c_{max} between a net pair. More specifically, if c_{max} is ten (units) and it is a three-layer routing solution S of t tracks, then in our proposed algorithm, our objective is to compute a three-layer routing solution S' of $t + k$ tracks for some minimum value of k (which is expected to be zero), in which the amount of crosstalk between a pair of nets is reduced to at most six units (or $0.6 \times c_{max}$) only. An outline of these devised algorithms is given below to see them at a glance.

Algorithm 1: Computation of Minimum Area Three-Layer Channel Routing Solutions

```
Procedure:  PRELIMINARY_3-LAYER_ROUTING_SOLUTION(C)

Input:      A channel instance C.

Output:     A three-layer minimum area routing solution S
            along with horizontal crosstalk.
```

1. Compute VCG and v_{max}, if the VCG is acyclic.

2. Compute HNCG, d_{max} (of the channel), and the maximum density zone(s) of the channel.

3. Compute oriented HNCG (OHNCG), where the edges are naturally oriented from left to right based on the presence of associated nets along the length of the channel.

4. **While** VCG is not null

    Consider the source vertex set in VCG.

    Compute weight of each of the source vertices using either 4.1 or 4.2.

4.1 **If** $\lceil d_{max}/2 \rceil > v_{max}$, **then**

        If a source vertex v_i passes through maximum density zone(s), **then**

            Weight = $\{ \Sigma zd(v_i) \} \times (\# \text{ maximum density zones})^P \}$ + $\Sigma ht(v_i)$

        Else

            Weight = $\Sigma zd(v_i) + \Sigma ht(v_i)$

        End If

    End If

4.2   **If** $\lceil d_{max}/2 \rceil \leq v_{max}$**, then**

> **If** a source vertex v_i in VCG has at least one path having height equal to v_{max}, **then**
>
> > Weight = {(Σheights of the source vertex) $\times$ (# paths with height v_{max})^q} + Σzd(v_i)
>
> **Else**
>
> > Weight = (Σheights of the source vertex) + Σzd(v_i)
>
> **End If**
>
> **End If**

4.3   Compute a maximum weighted clique $C_t(1)$ from OHNCG among the source vertex set in the current VCG.

4.4   Assign the intervals of the nets corresponding to the vertices in $C_t(1)$ to the current topmost tract t of the first horizontal layer of the channel, vertical segments of the assigned nets are also placed in the vertical layer of the channel.

4.5   Compute a subsequent maximum weighted clique $C_t(2)$ from OHNCG among the remaining source vertex set in the current VCG.

4.6   Assign the intervals of the nets corresponding to the vertices in $C_t(2)$ to the current topmost tract t of the second horizontal layer of the channel, vertical segments of the assigned nets are also placed in the vertical layer of the channel.

4.7   Delete all assigned nets to track t from HNCG and VCG.

4.8   Compute d_{max}, v_{max}, and maximum density zone(s) of the modified channel.

5.   **End While**

6.   Compute S along with its status of horizontal crosstalks.

End Procedure *PRELIMINARY_3-LAYER_ROUTING_SOLUTION(C)*

In this procedure, we have used some symbols that have been briefly explained as follows. Here for a source vertex v_i, $zd(v_i)$ and $ht(v_i)$ are the zonal density of the channel within the span of net n_i and the height of vertex v_i in the VCG, respectively [2]. Clearly, if this procedure iterates t times, then a t-track three-layer routing solution S for channel C is computed. Then we measure the amount of crosstalks of this initial routing solution S. In this work, this crosstalk is primarily the maximum amount of crosstalk between a pair of nets assigned to two adjacent tracks (in a horizontal layer) of S, as subsequently we are concerned in reducing crosstalk in terms of some bottleneck value (sacrificing more channel area, if at all necessary). An outline of the bottleneck crosstalk minimization algorithm is included in the subsequent section.

Algorithm 2: Computation of Minimum Crosstalk Three-Layer Channel Routing Solutions

As has already been mentioned, the three-layer routing solution S that is computed in the previous section may contain a maximum amount of overlapping (i.e., crosstalk in the horizontal layer) between a pair of nets beyond a limit, which may degrade the performance of the routing solution. In our next objective, we like to devise an algorithm in which the bottleneck value is set below the maximum amount of crosstalk obtained in S, and compute a routing solution that may contain an amount of crosstalk not crossing the bottleneck value (defined by us) in three-layer channel routing.

In our implementation, we like to reach the bottleneck crosstalk to a value not beyond 60% or 75% or 90% of the maximum value of crosstalk between a pair of nets that we obtained in S. Speculatively, the algorithm tries to move net intervals (in the horizontal layers) that are creating crosstalk beyond this limit satisfying all the constraints present in the channel and the assignment of nets to tracks in S. Groups of nets assigned to a track in S in two horizontal layers may also be exchanged. In doing all these, sometimes we may succeed, sometimes we may not. When it is a case of success, the routing area is not increased; otherwise, one more track is introduced into the channel (in S), and a desired bottleneck crosstalk routing solution S' is computed sacrificing channel area. In brief, the algorithm is given below to see it at a glance.

Procedure: *3-Layer_Bottleneck_Crosstalk_Reduction(S)*

Input: A three-layer routing solution S and a bottle-
                neck value x (in percent).

Output: Another three-layer routing solution S', in
                which the maximum crosstalk (for a pair of
                nets assigned to adjacent tracks in a horizon-
                tal layer) is at most x% of the maximum cross-
                talk (in a horizontal layer) in S.

1. Sort the responsible nets in non-ascending order P
    based on their maximum crosstalk with some other net
    and its presence from top to bottom in S.

2. **For** each such net n_j in P, perform either 2.1 or 2.2.

2.1 **If** n_j is isolated in the VCG, **then** it may be reassigned to a possibly different track in S, if a
suitable blank space is found in some other track
(in both the horizontal layers) where crosstalk is
reduced to at most $x\%$ of the maximum crosstalk in
S, satisfying horizontal constraints.

2.2 **If** n_j is in vertical constraint with n_i as well as n_k
such that (v_i, v_j) and (v_j, v_k) are edges in the VCG,
and n_i and/or n_k are there in P, while the nets n_i
and n_k are assigned to tracks t_i and t_k in S, respectively, $t_i < t_k$, **then** n_j may be reassigned to a possibly different track in S, if a suitable blank
space is found in some other track (in both the
horizontal layers) within the range of tracks t_i+1
and t_k-1, where crosstalk is reduced to at most $x\%$
of the maximum crosstalk in S, satisfying both horizontal as well as vertical constraints.

3. **If** no such track is found where n_j could be reassigned, **then** a blank track is introduced between the
tracks where n_j and the other responsible net(s) are
assigned (immediately above and/or below the track to
which n_j is assigned) that are creating crosstalk in S
beyond the allowable bottleneck (crosstalk) value.

4. Compute S', in which the maximum crosstalk is at most
$x\%$ of the maximum crosstalk in S.

End Procedure *3-Layer_Bottleneck_Crosstalk_Reduction(S)*

   Now, it is clear from the above algorithm (in Step 2.2), when v_j is a source (sink)
vertex, v_i (v_k) is no more there. In such a case, v_j may be attempted to reassign to
a track among the range of tracks 1 (t_i+1) and t_k-1 (the t-th track) in S (in both
the horizontal layers). Furthermore, if $t_k-t_i = 2$ and all the nets n_i, n_j, and n_k are in
P, then n_j is sandwiched in its track t_j, as $t_k-t_j = t_j-t_i = 1$. In such a case, we are
compelled to introduce blank tracks in between t_i and t_j and also in between t_j and
t_k (in both the horizontal layers). Of course, in such a case, in place of n_j, either
n_i or n_k (that renders more crosstalk with n_j in S) is better to be assumed earlier
for its reassignment to some other tracks in either of the horizontal layers. Inter-
estingly, these introduced blank tracks could be utilized for further reassignment of
nets, in any horizontal layer, for further reduction in horizontal crosstalk, if possible,
much below the bottleneck value, in order to compute S'. This completes the bottleneck crosstalk minimization algorithm starting from a three-layer given routing
solution S.

4.2 Computational Complexity

In algorithm *Preliminary_3-Layer_Routing_Solution*, different parameters and eventually the weights for each of the nets can be computed in $O(n)$ time as there are n nets in the given channel specification, C. The maximum weighted cliques in an iteration can be computed in time $O(n + e)$, where n and e are the order (which is also same as the number of nets in C) and size of the HNCG, respectively. As the algorithm iterates t times for computing a t-track routing solution S, the computational complexity that dominates the algorithm is $O(t(n + e))$. In this measurement of complexity, when $e = O(n)$, then t is also $O(n)$, and when $e = O(n^2)$, then t could be assumed as constant (as the number of tracks in computing S is reduced significantly).

On the other hand, when we measure crosstalk for the nets assigned to two adjacent tracks (in a horizontal layer), if there are $O(n)$ pairs of nets, this computation takes time $O(n)$. So, for a t-track routing solution S, the total crosstalk computation time may be overestimated as $O(nt)$. In the bottleneck crosstalk minimization algorithm *3-Layer_Bottleneck_Crosstalk_Reduction*, nets are expected to move to some other track, or to some other horizontal layer as well, if necessary, (satisfying constraints,) only when some suitable blank space is there in some other track for at least one responsible net creating more crosstalk in S. If no such track is found for such a desired interchange, a blank track is introduced in between (in both horizontal layers) eventually to make the earlier crosstalk negligible. This computation never exceeds the worst-case estimation of $O(n^2)$ for a given channel specification of n nets.

5 Experimental Results

To judge the novelty of the algorithms devised in this paper, we have randomly generated a huge number of general channel specifications in which the number of nets vary from 20 to 1000. More specifically, for each value of number of nets, we have randomly generated 200 general channel specifications each of which is free from any cyclic vertical constraint. The results we obtain are included in Table 1 following algorithm *Preliminary_3-Layer_Routing_Solution*. Along with crosstalk values on an average for 200 similar instances for a given value of net number (n), here we have comprised net wire length information as well of the computed routing solutions, since net wire length is also liable in high-performance routing in terms of delay.

Now, we compute reduced bottleneck crosstalk routing solutions, after algorithm *3-Layer_Bottleneck_Crosstalk_Reduction* over each of the three-layer minimum area channel routing solutions S (based on which the results are included in Table 1), and on an average the values of crosstalk we obtain for a given value of n are included in Tables 2, 3, and 4. In Table 2, the bottleneck value is set to 60% of the maximum amount of crosstalk obtained in S, whereas in Tables 3 and 4, the bottleneck values are set to 75% and 90%, respectively, of the maximum amounts of crosstalk obtained

Table 1 Three-layer routing solutions obtained after algorithm *Preliminary_3-Layer_Routing_Solution*

# Nets	Average d_{max}	Average v_{max}	Average # Tracks in S	Average horizontal net length	Average vertical net length	Average total net length	Average horizontal crosstalk	Average vertical crosstalk	Average total crosstalk
20	10.025	6.05	7.23	205.79	189.02	395.81	87.23	50.24	138.47
40	17.04	8.195	10.25	753.85	575.45	1330.2	414.15	168.14	583.29
60	24.39	9.885	14.36	1682.25	1166.22	2855.47	1041.17	352.56	1393.73
80	31.77	10.91	18.98	2973.02	1948.12	4921.14	1953.78	622.06	2576.84
100	38.36	12.35	21.23	4648.23	2910.07	7558.3	3180.56	932.12	4113.68
140	51.545	13.785	28.14	8819.19	5178.69	13997.88	6341.23	1689.12	8031.35
180	63.93	15.115	34.12	14299.31	8121.56	22421.87	10633.45	2680.15	13314.6
240	82.665	16.395	44.89	24921.02	13711.78	38631.8	19071.32	4552.35	23623.67
320	107.975	18.92	57.12	43929.15	23689.36	67618.51	34455.45	7950.2	42405.65
400	133.945	19.935	70.05	68342.04	36301.03	104643.07	54192.56	12207.3	66399.86
500	164.68	21.85	85.26	106304.28	55376.45	161681.73	85708.96	18734.54	104443.5
600	197.05	22.56	101.98	152328.29	78691.63	231019.92	124197.23	26756.32	150954.6
800	260.245	25.295	134.68	270081.29	138583.56	408664.85	222421.36	47272.09	269693.5
1000	321.75	27.085	164.12	418745.14	212492.89	631238.03	348524.65	72921.25	421445.9

Table 2 Three-layer reduced crosstalk routing solutions obtained after algorithm *3-Layer_Bottleneck_Crosstalk_Reduction* with 60% bottleneck values over the initial minimum area three-layer channel routing solutions

# Nets	Average d_{max}	Average v_{max}	Average # Tracks in S'	# Blank tracks introduced	Average horizontal crosstalk	% age reduction in horizontal crosstalk	Average vertical crosstalk	Average total crosstalk	% age reduction in total crosstalk
20	10.025	6.05	9.47	2.24	24.63	72.43	46.23	70.86	49.2
40	17.04	8.195	13.63	3.38	169.52	59.12	155.32	324.84	44.4
60	24.39	9.885	18.32	3.96	479.12	53.98	329.12	808.24	41.99
80	31.77	10.91	23.25	4.27	888.23	54.53	582.36	1470.59	42.93
100	38.36	12.35	27.12	5.89	1666.47	47.61	877.32	2544.79	38.14
140	51.545	13.785	35.62	7.48	3468.36	45.3	1598.32	5066.68	36.91
180	63.93	15.115	42.89	8.77	6117.87	42.47	2554.1	8671.97	34.87
240	82.665	16.395	53.23	8.34	11455.02	39.93	4343.78	15798.8	33.12
320	107.97	18.92	68.36	11.24	21712.03	36.98	7594.66	29306.69	30.89
400	133.94	19.93	93.14	14.58	44867.23	34.87	14760.78	59627.91	29.22
500	164.68	21.85	101.78	16.52	56765.26	33.76	18030.63	74795.89	28.38
600	197.05	22.56	119.26	17.28	83984.03	32.37	25804.63	109788.7	27.27
800	260.24	25.29	155.14	20.46	157074.7	29.38	45780.08	202854.8	24.78
1000	321.75	27.08	188.47	24.35	253295.5	27.32	70788.54	324084	23.1

Table 3 Three-layer reduced crosstalk routing solutions obtained after algorithm *3-Layer_Bottleneck_Crosstalk_Reduction* with 75% bottleneck values over the initial minimum area three-layer channel routing solutions

# Nets	Average d_{max}	Average v_{max}	Average # Tracks in S'	# Blank tracks introduced	Average horizontal crosstalk	% age reduction in horizontal crosstalk	Average vertical crosstalk	Average total crosstalk	% age reduction in total crosstalk
20	10.025	6.05	8.23	1	38.45	56.3	47.25	85.7	38.4
40	17.04	8.195	12.45	2.2	243.32	41.3	159.12	401.44	31.2
60	24.39	9.885	17.28	2.92	680.49	34.6	335.56	1016.15	27.13
80	31.77	10.91	21.17	2.19	1342.21	31.28	597.36	1939.57	24.72
100	38.36	12.35	24.04	2.81	2346.56	26.22	898.23	3244.79	21.12
140	51.545	13.785	30.95	2.81	4899.23	23.78	1637.36	6536.59	18.61
180	63.93	15.115	38.01	3.89	8512.14	19.94	2613.25	11125.39	16.44
240	82.665	16.395	48.3	3.41	15794.52	17.18	4446.23	20240.75	14.32
320	107.97	18.92	62.06	4.94	29094.23	15.55	7774.32	36868.55	13.05
400	133.94	19.935	75.26	5.21	46409.39	14.36	11985.14	58394.53	12.05
500	164.68	21.85	91.06	5.8	74612.04	12.94	18441.25	93053.29	10.9
600	197.05	22.56	108.23	6.25	109142.32	12.12	26344.24	135486.56	10.24
800	260.24	25.29	141.36	6.68	199352.23	10.37	46693.95	246046.28	8.76
1000	321.75	27.08	172.12	8	316649.24	9.14	72157.23	388806.47	7.74

Table 4 Three-layer reduced crosstalk routing solutions obtained after algorithm *3-Layer_Bottleneck_Crosstalk_Reduction* with 90% bottleneck values over the initial minimum area three-layer channel routing solutions

# Nets	Average d_{max}	Average v_{max}	Average # Tracks in S'	# Blank tracks introduced	Average horizontal crosstalk	%age reduction in horizontal crosstalk	Average vertical crosstalk	Average total crosstalk	%age reduction in total crosstalk
20	10.025	6.05	8	0.77	50.43	42.51	48.34	98.77	29
40	17.04	8.195	11.5	1.25	311.24	24.82	162.27	473.51	18.81
60	24.39	9.885	16.32	1.96	836.34	19.69	343.34	1179.68	15.36
80	31.77	10.91	20.46	1.48	1633.32	16.38	609.1	2242.42	12.96
100	38.36	12.35	23.56	2.33	2761.45	13.17	914.02	3675.47	10.64
140	51.545	13.785	30.34	2.2	5727.23	9.68	1672.32	7399.55	7.86
180	63.93	15.115	35.33	1.21	9860.1	7.26	2660.71	12520.81	5.96
240	82.665	16.395	46.54	1.65	17837.3	6.47	4510.2	22348.5	5.39
320	107.97	18.92	59.43	2.31	32621.23	5.32	7893.32	40514.55	4.45
400	133.94	19.935	72.23	2.18	51605.41	4.77	12135.23	63740.64	4
500	164.68	21.85	87.45	2.19	82261.1	4.02	18663.21	100924.31	3.36
600	197.05	22.56	103.65	1.67	119684.22	3.63	26635.32	146319.55	3.07
800	260.24	25.29	136.43	1.75	216014.12	2.88	47125.21	263139.33	2.43
1000	321.75	27.08	167.78	3.66	339625.23	2.55	72742.3	412367.53	2.15

in S. In each of these newly computed routing solutions S', additional tracks are also introduced in order to assure the bottleneck crosstalk value that has also been included in the relevant tables.

These tables also include the percentage reduction of horizontal crosstalk and the percentage reduction of total crosstalk. We may observe that while reducing horizontal crosstalk, vertical crosstalk may increase. However, as the channel length is much more on an average than the channel height, the total crosstalk is eventually reduced along with fulfilling the respective bottleneck values. As we are primarily concerned in attaining bottleneck crosstalk routing solutions, in these tables, we do not include the values against average net wire length obtained in these routing solutions.

Some graphical representations of the variation of bottleneck horizontal crosstalk against the number of nets are shown in Figs. 2 and 3. Figure 2 includes on an average bottleneck horizontal crosstalk versus number of nets when the bottleneck value is set to 60%, 75%, and 90% of the maximum amounts of horizontal crosstalk obtained in the minimum area three-layer routing solutions after algorithm *Preliminary_3-Layer_Rou-ting_Solution*. On the other hand, Fig. 3 shows the average percentage reduction in bottleneck horizontal crosstalk after algorithm *3-Layer_Bottleneck_Crosstalk_Reduction* against number of nets when the bottleneck value is set to 60%, 75%, and 90% of the maximum amounts of crosstalk obtained in the initial minimum area three-layer channel routing solutions.

Figure 4 includes only four hardcopy routing solutions that have been computed based on the implementation of our proposed heuristic algorithms. None of the minimum area routing solutions after algorithm *Preliminary_3-Layer_Routing_Solution* have shown here. However, only four (out of 2800) routing solutions after the *3-Layer_Bottleneck_Crosstalk_Reduction* algorithm have been depicted, out of which

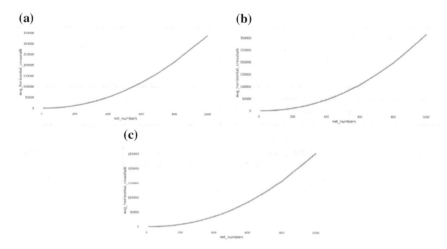

Fig. 2 Bottleneck horizontal crosstalk against number of nets when the bottleneck value is set to **a** 60%, **b** 75%, and **c** 90% of the maximum amounts of crosstalk obtained in the minimum area three-layer routing solutions after algorithm *Preliminary_3-Layer_Routing_Solution* on an average

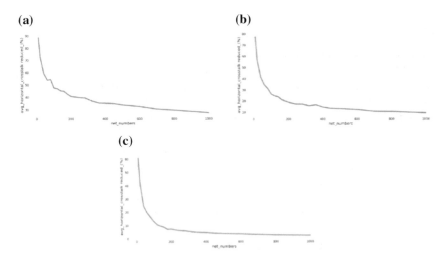

Fig. 3 Average percentage reduction in bottleneck horizontal crosstalk after algorithm *3-Layer _Bottleneck_Crosstalk_Reduction* against number of nets when the bottleneck values are set to **a** 60%, **b** 75%, and **c** 90% of the maximum amounts of crosstalk obtained in the initial minimum area three-layer channel routing solutions

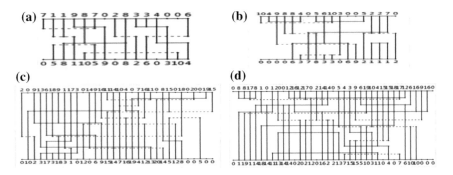

Fig. 4 A set of three-layer routing solutions computed with reduced crosstalk after algorithm *3-Layer_Bottleneck_Crosstalk_Reduction* against numbers of nets, in which a relevant bottleneck value is set based on the maximum amounts of crosstalk obtained in the initial minimum area channel routing solutions after algorithm *Preliminary_3-Layer_Routing_Solution*. **a** Number of nets is 10 and the bottleneck value is set to 2 for each of the layers; no blank track has been introduced. **b** Number of nets is 10 and the bottleneck value is set to 3 for each of the layers; three blank tracks have been introduced. **c** Number of nets is 20 and the bottleneck value is set to 7 for each of the layers; one blank track has been introduced. **d** Number of nets is 20 and the bottleneck value is set to 11 for each of the layers; no blank track has been introduced

no additional (blank) track is introduced to achieve the calculated bottleneck value for each of the routing solutions in (a) and (d), whereas for the routing solution in (b) and (c), respectively, three and one additional blank track have been introduced to satisfy the bottleneck value set for reducing the crosstalk.

6 Conclusion

In this paper, we have considered the bottleneck crosstalk minimization problem in three-layer channel routing. We know that the area minimization problem in the reserved three-layer Manhattan channel routing model is NP-hard. The defined crosstalk minimization problem is also known to be NP-hard. Thus, devising heuristic algorithms is a natural choice, hopefully, to compute expected routing solutions. In this paper, we have designed and implemented two such heuristic algorithms that are executed one after the other. A large class of random channel instances has been generated for analyzing the performance of the devised algorithms. Our crosstalk reduction algorithm is implemented for computing routing solutions with 60, 75, and 90% bottleneck consideration over the maximum amount of crosstalk present in an initial minimum area three-layer channel routing solution. All these experimental results have been presented in this paper and a few hardcopy routing solutions have also been exemplified. As an immediate enhancement over the work, we may think of bottleneck crosstalk minimization in multilayer channel routing.

References

1. Sherwani, N.A.: Algorithms for VLSI Physical Design Automation. Kluwer Academic Publishers, Boston (1993)
2. Pal, R.K.: Multi-Layer Channel Routing: Complexity and Algorithms, Narosa Publishing House, New Delhi (Also published from CRC Press, Boca Raton. USA and Alpha Science International Ltd., UK) (2000)
3. Yoshimura, T., Kuh, E.S.: Efficient algorithms for channel routing. IEEE Trans. CAD of Integr. Circuits Syst. **1**, 25–35 (1982)
4. Gao, T., Liu, C.L.: Minimum crosstalk channel routing. In: Proceedings of IEEE International Conference on Computer-Aided Design, pp. 692–696 (1993)
5. Pal, A., Chaudhuri, A., Pal, R.K., Datta, A.K.: Hardness of crosstalk minimisation in two-layer channel routing. Integr. VLSI J. (Elsevier) (ISSN: 0167-9260), **56**, 139–147 (2017)
6. Mandal, T.N., Mehera, R., Datta, A.K., Pal, R.K.: Hardness of crosstalk minimisation in three-layer channel routing. Manuscript (2019)
7. Mandal, T.N., Dutta Banik, A., Dey, K., Mehera, R., Pal, R.K.: Algorithms for minimizing bottleneck crosstalk in two-layer channel routing. In: Presented in the 2nd International Conference on Computational Advancement in Communication Circuit and System (ICCACCS 2018) held in Kolkata, India during November 23–24 (2018)
8. Golumbic, M.C.: Algorithmic Graph Theory and Perfect Graphs. Academic Press, New York (1980)

9. Pal, R.K., Datta, A.K., Pal, S.P., Pal, A.: Resolving horizontal constraints and minimizing net wire length for VHV channel routing. Technical Report: TR/IIT/CSE/92/01, Department of Computer Science and Engineering, IIT, Kharagpur (1992)

10. Hashimoto, A., Stevens, J.: Wire routing by optimizing channel assignment within large apertures. In: Proceedings of the 8th ACM Design Automation Workshop, pp. 155–169 (1971)

11. Pal, R.K., Datta, A.K., Pal, S.P., Pal, A.: Resolving horizontal constraints and minimizing net wire length for multi-layer channel routing. In: Proceedings of IEEE Region 10's Eighth Annual International Conference on Computer, Communication, Control and Engineering (TENCON 1993), vol. 1, pp. 569–573 (1993)

12. Pal, R.K., Datta, A.K., Pal, S.P., Das, M.M., Pal, A.: A general graph theoretic framework for multi-layer channel routing. In: Proceedings of the Eighth VSI/IEEE International Conference on VLSI Design, pp. 202–207, Jan. 4–7, 1995

13. Pal, A., Kundu, D., Datta, A.K., Mandal, T.N., Pal, R.K.: Algorithms for reducing crosstalk in two-layer channel routing. J. Phys. Sci. **10**, 167–177, Dec. 2006 (ISSN: 0972-8791)

14. Pal, A., Mandal, T.N., Khan, A., Pal, R.K., Datta, A.K., Chaudhuri, A.: Two algorithms for minimizing crosstalk in two-layer channel routing. Int. J. Emer. Trends Technol. Comput. Sci. (IJETTCS) **3**(6), 194–204 (2014) (ISSN: 2278-6856)

15. Schaper, G.A.: Multi-layer channel routing, Ph.D. Thesis, Department of Computer Science, University of Central Florida, Orlando (1989)

Arithmetic Circuits Using Reversible Logic: A Survey Report

Arindam Banerjee and Debesh Kumar Das

Abstract In this paper, a survey has been made on the design of arithmetic circuits like adder, subtractor, multiplier, and squarer. There are many design schemes for those arithmetic circuits some of which have been studied and described in this paper.

Keywords Adder · Subtractor · Multiplier · Squarer · Reversible logic

1 Introduction

Arithmetic circuits are the core components of ALU design. Dedicated arithmetic circuits have the advantages of executing fast operation. Thus, a lot of researchers concentrate on designing dedicated arithmetic circuits. Among the arithmetic circuits, adder, subtractor, multiplier, squarer, etc. are most common and many researchers have worked on different synthesis techniques to design these circuits. In conventional technology, power consumption of these complex circuits is alarmingly large. Thus, from decades ago, the researchers try to find out alternative solution to reduce power consumption.

As one of the promising alternatives to reduce power consumption, reversible computation has been introduced as the pathfinder. Reversible logic is by nature bijective, i.e., it employs n-input n-output functions that map each input vector to a unique output vector. It allows to derive inputs from outputs and vice versa, i.e., all computations can be reverted. The reversibility provides certain aspects in low power design. The logic provides lossless information processing. In 1961, Landauer [1] postulated that energy must be dissipated as heat only when information is destroyed. More precisely, the minimum amount of energy that is dissipated for each lost bit of

A. Banerjee (✉)
Department of Electronics and Communication Engineering, JIS College of Engineering, Kalyani, Nadia, West Bengal, India
e-mail: banerjee.arindam1@gmail.com

D. K. Das
Department of Computer Science, Jadavpur University, Kolkata, Jadavpur, India

© Springer Nature Singapore Pte Ltd. 2020
R. Chaki et al. (eds.), *Advanced Computing and Systems for Security*,
Advances in Intelligent Systems and Computing 995,
https://doi.org/10.1007/978-981-13-8962-7_8

information is related to a quantity $KT ln(2)$ $Joule$ [2] (where K is the Boltzmann Constant and T is the temperature in Kelvin scale). In 1973, Bennett [3] has proposed that the above heat dissipation can be reduced if no information is destroyed which ultimately requires reversible logic. These claims have been verified experimentally and reported in [4]. Thus, as an indispensable alternative to transistor or CMOS technology, reversible logic has been introduced. It has a wide scope of application in nanotechnology and quantum computing.

Therefore, the synthesis of the arithmetic functions has drawn the attention of many researchers in the past and at present both in conventional [5–11] and reversible technology [12–17].

2 Reversible Logic

As per [18], *A multi-output function is reversible if and only if its number of inputs* *(n) is equal to the number of outputs (m) (i.e., n = m) and it maps each input pattern* *to a unique pattern.* In other words, a reversible Boolean function is a bijection that performs a permutation to the set of input patterns. A Boolean function which is not reversible is termed as irreversible function. In reversible logic, there are different terms like ancillary input, garbage output, and quantum cost. Ancillary inputs are the extra inputs other than the primary inputs having constant values to design reversible circuits. Garbage outputs are extra outputs other than primary outputs to maintain logical reversibility. Quantum cost is defined as the number of elementary quantum operations required for a reversible circuit.

3 Survey on Different Arithmetic Circuits Design in Reversible Logic

There are many literature on arithmetic circuits like adder, subtractor, multiplier, squarer, etc. in reversible logic. Few designs have been studied and described below.

3.1 Adder and Subtractor Design in Reversible Logic

The first adder design in reversible logic has been implemented using Peres gate [19]. Figure 1 shows the adder design using Peres gate. In Fig. 1, a and b are control inputs (for adder, they are the operands) and t is the target input. $Sum = a \oplus b$ and $Cout = t \oplus ab$. If $t = 0$ then $Cout = ab$, i.e., the carry output. This is the design of a half adder.

Fig. 1 Adder design using
Peres gate

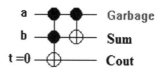

Fig. 2 Adder design using
Double Peres gate

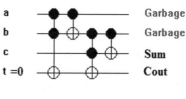

Fig. 3 Full adder proposed
by Cuccaro [20]

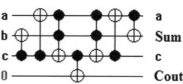

Full adder design has been achieved using double Peres gate as shown in Fig. 2. Here a, b, and c are the control inputs (for adder design they are the operands and particularly c is the previous carry) and t is the target input. $Sum = a \oplus b \oplus c$ and $Cout = t \oplus ab \oplus (a \oplus b)c$. If $t = 0$ then $Cout = ab \oplus (a \oplus b)c$, i.e., the final carry output.

Another full adder design has been proposed in [20] which has been shown in Fig. 3. Here the constant 0 is the target input. The following equations show the Boolean expressions for *Sum* and *Cout*.

$$Cout = ab \oplus (a \oplus b)c \tag{1}$$

$$= c \oplus c \oplus c(a \oplus b) \oplus ab \tag{2}$$

$$= c \oplus (c \oplus a)(c \oplus b) \tag{3}$$

At first, carry is generated and then from the carry sum is generated from Eq. 3 as shown below:

$$Sum = a \oplus b \oplus c \tag{4}$$

$$= Cout \oplus (c \oplus a)(c \oplus b) \oplus (c \oplus a) \oplus (c \oplus b) \tag{5}$$

There is another design for addition which has been proposed in [14]. The design is for both binary and BCD number systems. In binary number system, the following expressions have been used:

$$Sum = a \oplus b \oplus c \tag{6}$$

$$Cout = z \oplus ab \oplus (a \oplus b)c = z \oplus ab \oplus ac \oplus bc \tag{7}$$

Fig. 4 TR gate proposed in
[22]

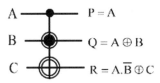

Here z is the target input which is 0 in this case to achieve final carry as shown in
[14]. Here the N-bit design has been achieved using single ancillary input.

Like adder, there are few designs for subtraction [21, 22]. The design proposed in
[21] is for BCD subtraction, whereas in [22] binary subtraction has been reported. In
[21], the subtraction has been achieved using $9's$ complement and parallel addition
technique. Here addition has been implemented using both ripple carry and carry
look-ahead adders but in each adder cell, full adder has been replaced by modified
full adder which has been shown in [21]. In [22], the subtractor design has been
achieved using TR gate which is shown in Fig. 4.

3.1.1 Comparative Study of Adder Architectures

A comparative study for reversible adder is shown here. In the literature [14], it has
been claimed that the design is garbage free. But if we consider only the primary outputs then the design produces some garbage outputs. The details of the comparative
study is given in Table 1.

3.2 Multiplier Design in Reversible Logic

There are many literature on multiplier design in reversible logic. One design for
multiplication has been proposed in [23] using Fredkin gate [24] and TSG gate
shown in Fig. 5a, b.

Another multiplier in reversible logic has been proposed in [25]. Here array multiplier technique has been adopted and implemented using reversible gates.

Another multiplier design using the well-known Karatsuba's [26] technique has
been proposed in [27]. In this technique, operands are partitioned into two equal

Table 1 Comparative study of the adder structures for n-bits

Parameters	[19]	[20]	[14]
Ancillary inputs	n	1	1
Garbage outputs	$2n - 1$	n	n
Quantum cost	$6n$	$12n + 1$	$13n - 10$

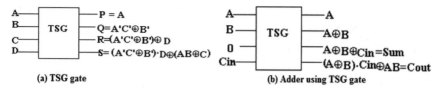

(a) TSG gate   (b) Adder using TSG gate

Fig. 5 **a** TSG gate and **b** adder using TSG gate

```
1    if  (n < turningPoint)
2         c = MULT_H (a, b)
3
4    if  (n%2 = 1)
5         init  a_n, b_n, c_{2·n}, c_{2·n+1}  with  0
6
7    k := ⌊n/2⌋
8    init  d, e  (k+1 bits),  f  (2·k+2 bits)  with  0
9    c̄ = MULT_K(ā, b̄)
10   c = MULT_K (a, b)
11   d = ā + a
12   e = b̄ + b
13   h = MULT_K(d, e)
14   h− = c̄
15   h− = c
16   T_{i=k}^{3·k+3} c_i+ = h
```

Fig. 6 Algorithm of Karatsuba method for multiplication

Fig. 7 Technique for
constant integer
multiplication

parts, higher order and lower order and crosswise multiplication technique has been adopted. Figure 6 shows the Karatsuba.s method of multiplication.

Garbage-free integer multiplication technique with constants has been reported in [28]. Here one of the operands is a constant integer of the type $(2^n \pm 2^l \pm 1)$ and the other operand is any integer variable. The technique is shown in Fig. 7. For example, shown in Fig. 7, if an operand A is multiplied by $5 = 2^2 + 1$ then it can be achieved by multiplying A by 4 which is basically a 2 bit left shifting operation and then adding the operand.

Another multiplier design in reversible logic has been reported in [29] using ancient Indian mathematics. The design is efficient for a particular range of operands in close vicinity to a particular radix 100 (8 bit representation). The technique used in [29] is known as "Nikhilam" rule. Figure 8a shows the algorithm for multiplication using "Nikhilam" rule. Here X and Y are the operands and the radix is taken to be 100. Figure 8b shows the schematic diagram for the proposed multiplier.

Algorithm.
step1: Subtract inputs X,Y from 100,
    (t1=100-X, t2=100-Y).
step2: Multiply **t1** and **t2**, (**PL=t1*t2**).
step3: Subtract **t2** from **X**, (**PH =X-t2**).
step3: Merge **PH** and **PL**,(**Out={PH,PL}**).

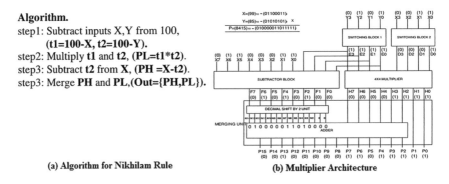

(a) **Algorithm for Nikhilam Rule**   (b) **Multiplier Architecture**

Fig. 8 **a** Algorithm for multiplication, **b** Corresponding architecture as described in [29]

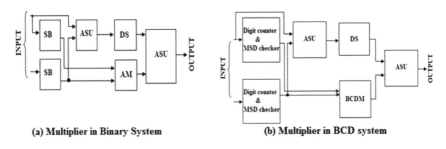

(a) **Multiplier in Binary System**   (b) **Multiplier in BCD system**

Fig. 9 Schematic diagram for reversible **a** binary, **b** BCD multiplier as described in [30]

Using ancient Indian mathematics, another efficient 8-bit multiplier in reversible logic has been proposed in [30] for both binary and decimal number system using the "Nikhilam" as discussed earlier. For this technique, the range of the operands ranges between $radix \pm 15$. Therefore, here the operand size is not restricted to 8 bit only, it may be greater than that. Figure 9a shows the schematic diagram of the proposed multiplier for binary number system. Here "SB", "ASU", "DS", and "AM", respectively, indicate the switching block, adder–subtractor unit, decimal shifter, and the array multiplier. In [30], another multiplier design has been shown for decimal number system. Figure 9b shows the schematic diagram for decimal multiplier. In Fig. 9b, the architecture contains the blocks MSD (Most significant Digit) checker and BCD (Binary Coded Decimal) multiplier.

Another multiplication technique in reversible logic has been reported in [13] using binary tree optimization technique. Figure 10a shows the binary tree representation for $N \times N$ multiplier where N is the number of bits. In Fig. 10a, PS indicates the partial sum which is basically generated from the partial products in the multiplication scheme. Figure 10b shows the schematic diagram of the proposed 4×4 reversible multiplier.

A low power multiplier architecture in reversible logic has been proposed in [16]. In this design, for addition, two architectures have been proposed using Peres gate and TSG gate which have been shown in Fig. 5. 4×4 bit multiplier design has been shown in [16].

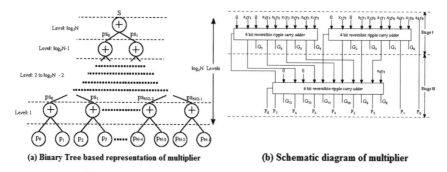

(a) Binary Tree based representation of multiplier   **(b) Schematic diagram of multiplier**

Fig. 10 **a** Binary tree representation of $N \times N$ multiplier, **b** Schematic diagram of multiplier

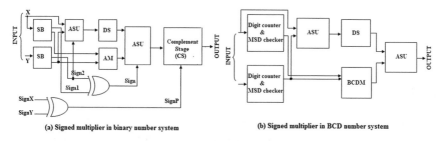

(a) Signed multiplier in binary number system   **(b) Signed multiplier in BCD number system**

Fig. 11 Schematic diagram for reversible **a** binary multiplier, **b** BCD multiplier as shown in [31]

Another hardware-efficient reversible multiplication technique for signed number system has been proposed in [31] using the ancient Indian mathematics for both binary and decimal number system. The same "Nikhilam" rule as discussed earlier has been adopted in this scheme. Figure 11a shows the architecture for binary multiplier for signed number system. Here $Sign X$ and $Sign Y$ are the signs of the operands X and Y and "CS" indicates the complement stage which is required if the result is negative, i.e., $2's$ complemented. Figure 11b shows the architecture for decimal multiplier for signed number system which is identical to Fig. 9b because the sign bits do not affect the result of the last ASU.

Another multiplication scheme in reversible logic has been proposed in [17] using systolic array based technique. Figure 12 shows the structure for systolic array multiplication for three bits. Here "PE" is the processing element. For multiplier, processing element consists of basically AND gate and adder circuit. Here $a_{11}, a_{12}, a_{13}, \ldots, a_{32}, a_{33}$ and $b_{11}, b_{12}, b_{13}, \ldots, b_{32}, b_{33}$ are input operands. The placement of the bits is aligned as shown in Fig. 12. The reversible implementation has been achieved by constructing the processing elements by Toffoli gates and Peres gates followed by shift registers. In [17], the shift register has also been designed in reversible logic. Moreover, ASIC implementation of the design has been shown in [17].

Fig. 12 Schematic diagram
of systolic array structure

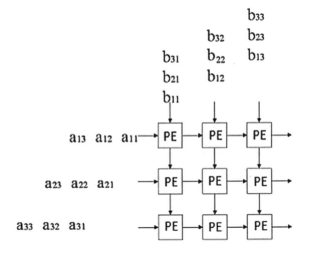

3.2.1 Comparison of Different Multipliers

The reversible multiplier designs shown [16, 23, 25] are the array multiplication techniques for 4-bit only, whereas, in [27], a generic multiplier architecture has been discussed. The parameters of the reversible multipliers [23, 25, 27, 29–31] have been calculated and shown in Tables 2 and 3. Except the design shown in [31], all the other designs compared here are for unsigned numbers.

From Table 3, it is noticed that for BCD number system, signed operation does not affect the original unsigned multiplication and thus the results shown in columns 5 and 7 are identical. For the technique shown in [28], one operand is constant and is of a particular format ($2^n \pm 2^l \pm 1$) which is not generic to be considered for comparison.

Table 2 Comparative study of reversible multipliers for 4 bits

Parameters	[23]	[25]	[27]
Ancillary inputs	18	28	16
Garbage outputs	26	28	12
Quantum cost	78	137	518

Table 3 Parametric computation of unsigned and signed reversible multipliers for 8 bits (For BCD system, 2 digits)

Parameters	[27]	[29]	[30] (Binary) (Unsigned)	[30] (BCD) (Unsigned)	[31] (Binary) (Signed)	[31] (BCD) (Signed)
Ancillary inputs	54	83	51	57	57	57
Garbage outputs	46	94	55	90	61	90
Quantum cost	2437	458	467	521	507	521

3.3 Squarer Design in Reversible Logic

Like multiplier, there are few designs on square computations [12, 32, 33] in reversible logic. The squarer design described in [12] is based on the array multiplication technique. The overall design has been achieved using Toffoli, Peres, and double Peres gates.

In [32], another squarer design has been proposed in reversible logic. Here iterative structure has been used to reduce the line count and quantum cost. Figure 13 shows the structure from which it is obvious that N-bit squarer design has been achieved using $(N - 1)$-bit squarer design. The design scheme has been implemented using two types of adder (i) using Peres gate ($Technique_1$) and (ii) the design proposed in [20] ($Technique_2$).

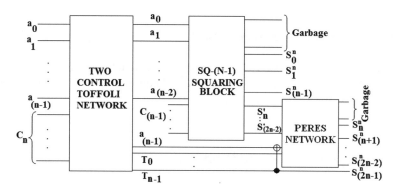

Fig. 13 Schematic diagram for n bit squaring

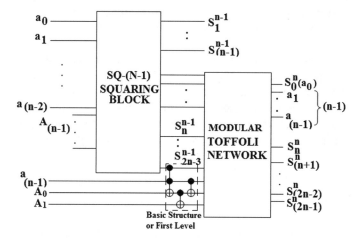

Fig. 14 Schematic Diagram for n-bit squaring

Table 4 Comparative study of reversible squarer for 16 bits

Parameters	[12]	[32] ($Technique_1$)	[32] ($Technique_2$)	[33]
Ancillary inputs	241	223	212	31
Garbage outputs	225	206	195	0
Quantum cost	1290	792	865	1861

In [33], another squarer design has been proposed in reversible logic which is also based on iteration. Here a new addition technique has been implemented using Toffoli gates only. Figure 14 shows the architecture where the adders have been implemented using the modular Toffoli network. Moreover, an efficient gate elimination algorithm has been proposed to eliminate few redundant Toffoli gates.

3.3.1 Comparison of Different Squarer

The reversible squarer reported in [12] is based on array multiplier structure, whereas the designs reported in [32, 33] are based on iterative structure. The parametric comparison has been shown in Table 4.

4 Conclusion and Future Direction

In this paper, many arithmetic circuit design techniques for adder, subtractor, multiplier, and squarer have been shown. Though the designs have optimized hardware overhead still more optimized design should be required in future keeping track with the more complex operations required gradually day by day.

Since there are several designs on reversible adder, subtractor, multiplier, and squarer, future research should be directed to the design of other arithmetic circuits like divider, square root, exponentiation, logarithm, etc. using these circuits which are best known so far. Moreover, research emphasis can be given to design DSP architectures like discrete Fourier transform, fast Fourier transform, convolution, digital filters using different emerging technologies like reversible logic, memristors, optical computation technique, quantum cellular automata (QCA), etc.

References

1. Landauer, R.: Irreversibility and heat generation in the computing process. IBM J. Res. Dev. **5**, 183 (1961)
2. Castello, D.J., Forney, G.D.: Cannel coding: the road to cannel capacity. Proc. IEEE. **95**(6), 1150–1177 (2007)

3. Bennett, C.H.: Logical reversibility of computation. IBM J. Res. Dev. **17**(6), 525–532 (1973)
4. Berut, A., Arakelyan, A., Petrosyan, A., Ciliberto, S., Dillenschneider, R., Lutz, E.: Experimental verification of landauer/'s principle linking information and thermodynamics. Nature **483**(7388), 187–189 (2012)
5. Hong, S., Kim, S., Papaefthymiou, M.C., Stark, W.E.: Low power parallel multiplier design for dsp applications through co-efficient optimization. In: IEEE International Conference on ASIC/SOC, pp. 286–290 (1999)
6. Bulic, P., Babic, Z., Avramovic, A.: A simple pipelined logarithmic multiplier. In: IEEE International Conference on Computer Design, pp. 235–240, October 2010
7. Mrazek, V., Sarwar, S.S., Sekanina, L., Vasicek, Z., Roy, K.: Design of power-efficient approximate multipliers for approximate artificial neural networks. In: IEEE/ACM International Conference on Computer-Aided Design, pp. 1–7 (2016)
8. Venkatachalam, S., Ko, S.B.: Design of power and area efficient approximate multipliers. In: IEEE Transactions on Very Large Scale Integration (VLSI) Systems, vol. 25(5), pp. 1782–1786 (2017)
9. Yoo, J.T., Smith, K.F., Gopalakrishnan, G.: A fast parallel squarer based on divide-and-conquer. IEEE J. Solid-State Circuits **32**, 909912 (June 1997)
10. Deshpande, A., Draper, J.: Comparing squaring and cubing units with multipliers. In: IEEE 55th International Midwest Symposium on Circuits and Systems (MWSCAS), pp. 466–469 (2012)
11. Datla, S.R., Thornton, M.A., Matula, D.W.: A low power high performance radix-4 approximate squaring circuit. In: 20th IEEE International Conference on Application Specific Systems, Architectures and Processors (ASAP), Vol. 7, pp. 91–97 (July 2009)
12. Jayashree, H.V., Thapliyal, H., Agrawal, V.K.: Design of dedicated reversible quantum circuitry for square computation. In: Proceedings of 27th International Conference on VLSI Design, pp. 551–556 (January 2014)
13. Kotiyal, S., Thapliyal, H., Ranganathan, N.: Circuit for reversible quantum multiplier based on binary tree optimizing ancilla and garbage bits. In: Proceedings of 27th International Conferenceon VLSI Design, pp. 545–550 (January 2014)
14. Thapliyal, H., Ranganathan, N.: Design of efficient reversible logic based binary and bcd adder circuits. ACM J. Emerging Technol. Comput. Syst. **9**(3), 17:1–17:31 (September 2013)
15. Thapliyal, H., Ranganathan, N., Kotiyal, S.: Design of testable reversible sequential circuits. IEEE Trans. VLSI **21**(7), 1201–1209 (2013)
16. Thakre, A.K., Chiwande, S.S., Chafale, S.D.: Design of low power multiplier using reversible logic gate. In: Proceedings of International Conference on Green Computing Communication and Electrical Engineering (6–8 March 2014)
17. Madhulika, C., Nayak, V.S.P., Prasanth, C., Praveen, T.H.S.: Design of systolic array multiplier circuit using reversible logic. In: 2017 2nd IEEE International Conference on Recent Trends in Electronics, Information and Communication Technology, pp. 1670–1673 (2017)
18. Wille, R., Drechsler, R.: Towards a Design Flow for Reversible Logic. Springer (2010)
19. Peres, A.: Reversible logic and quantum computers. APS Phys. Rev. A **32**, 3266–3276 (1985)
20. Cuccaro, S.A., Draper, T.G., Kutin, S.A.: A new quantum ripple-carry addition circuit. arXiv:quant-ph/0410184. (February 2008)
21. Thapliyal, H., Arabnia, H., Srinivas, M.: Efficient reversible logic design of bcd subtractors. Springer Trans. Comput. Sci. J. **3**(LNCS 5300), 99–121 (2009)
22. Thapliyal, H., Ranganathan, N.: A new design of the reversible subtractor circuit. In: Proceedings of the 11th IEEE International Conference on Nanotechnology (IEEE NANO), pp. 1430–1435 (August 2011)
23. Thapliyal, H., Srinivas, M.B.: Novel reversible multiplier architecture using reversible tsg gate. In: IEEE International Conference on Computer Systems and Applications, 8th March 2006
24. Fredkin, E.F., Toffoli, T.: Conservative logic. Int. J. Theor. Phys. **21**(3), 219–253 (1982)
25. Haghparast, M., Jassbi, S., Navi, K., Eshghi, M.: Optimized reversible multiplier circuits. J. Circuits Syst. Comput. **18**, 311–323 (2009)

26. Karatsuba, A., Ofman, Y.: Multiplication of many-digital numbers by automatic computers. Doklady Akad. Nauk SSSR **145** (1963)
27. Offermann, S., Wille, R., Dueck, G.W., Drechsler, R.: Synthesizing multiplier in reversible logic. In: 13th IEEE Symposium on DDECS, pp. 335–340 (April 2010)
28. Axelsen, H.B., Thomsen, M.K.: Garbage-free integer multiplication with constants of the form $2^k \pm 2^l \pm 1$. In: 4th Workshop on Reversible Computation (July 2012)
29. Saravanan, P., Chadrasekar, P., Chandran, L., Sriram, N., Kalpana, P.: Design and implementation of efficient vedic multiplier using reversible logic. In: International Symposium on VLSI Design and Test, pp. 364–366 (2012)
30. Banerjee, A., Das, D.K.: The design of reversible multiplier using ancient indian mathematics. In: International Symposium on Electronic Design, pp. 31–35 (December 2013)
31. Banerjee, A., Das, D.K.: The design of reversible signed multiplier using ancient indian mathematics. J. Low Power Electron. **11**, 467–478 (December 2015)
32. Banerjee, A., Das, D.K.: Squaring in reversible logic using iterative structure. In: Proceedings of East West Design and Test Symposium (September 2014)
33. Banerjee, A., Das, D.K.: Squaring in reversible logic using zero garbage and reduced ancillary inputs. In: International Conference on VLSI Design (2015)

Author Index

© Springer Nature Singapore Pte Ltd. 2020
R. Chaki et al. (eds.), *Advanced Computing and Systems for Security*,
Advances in Intelligent Systems and Computing 995,
https://doi.org/10.1007/978-981-13-8962-7

Printed in the United States
By Bookmasters